ADVANCED CERAMIC MATRIX COMPOSITES

ADVANCED CERAMIC MATRIX COMPOSITES

Design Approaches, Testing and Life Prediction Methods

Edited by

EDWARD R. GENERAZIO, Ph.D.

National Aeronautics and Space Administration
Langley Research Center
Hampton, Virginia, U.S.A.

CRC Press
Taylor & Francis Group
Boca Raton London New York

CRC Press is an imprint of the
Taylor & Francis Group, an **informa** business

Table of Contents

Part 3: Nondestructive Evaluation

Preface

A DVANCED high-temperature, low-density composite materials are being developed for use as propulsion and structural components in the next generation of aerospace systems. Transition-to-practice technologies are technologies needed to bring brittle matrix composites up to the level necessary for practical application as aerospace materials. The primary technologies are mechanical testing, nondestructive evaluation (NDE), mechanical design, and life prediction. Manufacturing is also a key area for which there is currently little information for brittle matrix composites. It is expected that the manufacturing area will grow and will become a critical factor in determining the return on investment for the brittle system.

Two key areas where technologies are being used to develop brittle matrix composites are materials characterization and component design. Materials characterization techniques, both destructive and nondestructive, yield information on material failure mechanisms, integrity, and degradation. This crucial information guides and enhances the development of brittle matrix composites. Advanced design tools are also being developed to assist component designers with structural brittle matrix composite components by determining the component's reliability and life expectancy. Two types of NDE are needed for developing brittle matrix composite components: *in situ* and in-process NDE for process monitoring and control. In-process NDE is done during the development of materials to characterize microstructural and flaw populations that affect strength and life. *In situ* NDE is performed during testing and is used for identifying the occurrence of failure and the failure mechanisms. Both types of NDE provide feedback for materials development and for validating analytical models.

The mechanical failure of brittle materials follows probabilistic failure theories; therefore, probabilistic methods need to be incorporated into the design when developing a specific component. Probabilistic analytical methods and computer codes are being developed for predicting fast fracture and life remaining of laminated composite materials. The overall system level benefits need to be determined a priori before embarking on a path to incorporate brittle matrix materials into a design. One such system, Technology Benefit Estimator (T/BEST), has been developed for evaluating the benefits of introducing new technologies into aerospace systems. The benefits obtained from using a formal method to assess benefits of new technologies cannot be understated.

EDWARD R. GENERAZIO

PART 1

DESIGNING

CHAPTER 1

Designing Ceramic Components for Durability

NOEL N. NEMETH,[1] LYNN M. POWERS,[2]
LESLEY A. JANOSIK[1] and JOHN P. GYEKENYESI[1]

OVERVIEW

SUCCESSFUL application of advanced ceramics depends on proper characterization of material properties and the use of a probabilistic brittle material design methodology. Described here is a computer program for predicting the probability of a monolithic ceramic component's failure as a function of its time in service.

The increasing importance of ceramics as structural materials places a high demand on assuring component integrity while simultaneously optimizing performance and cost. High-temperature structural applications include heat exchangers and spark ignition, diesel, and turbine engine components. Other uses include wear parts (nozzles, valves, seals, etc.), cutting tools, grinding wheels, bearings, coatings, electronics, and human prostheses. Designing ceramic components for durability requires that all potential failure modes are identified and accounted for in a comprehensive design methodology.

Ceramic properties have progressively improved due to advances in both processing and composition. This has reduced the number and size of defects, led to the development of tougher materials that better tolerate the presence of flaws, and improved understanding and control of microstructural composition. However, ceramics are inherently brittle, and the lack of ductility leads to low strain tolerance, low fracture toughness, and large variations in observed fracture strength caused by the variable severity (size) and random distribution of flaws.

[1]NASA Lewis Research Center, Cleveland, OH, U.S.A.
[2]Department of Civil Engineering, Cleveland State University, Cleveland, OH, U.S.A.

The ability of a ceramic component to sustain load degrades over time as a result of oxidation, creep, stress corrosion cracking, and elevated-temperature use. The temperature and load conditions where these failure modes are active are illustrated using fracture mechanism maps. Stress corrosion cracking, cyclic fatigue, and elevated-temperature sustained load response are different aspects of the phenomenon called subcritical crack growth (SCG).

SCG refers to the progressive extension of a crack over time. An existing flaw extends until it reaches a critical length, causing catastrophic crack propagation. Under the same conditions of temperature and load, ceramic components display large variations in rupture times from SCG. Unfortunately, the small critical flaw size and large number of flaws make it difficult to detect beforehand the particular flaw that will initiate component failure. Probabilistic design techniques are therefore necessary to predict the probability of a ceramic component's failure as a function of service time.

In the United States, development of brittle material design technology has been actively pursued during the past two decades. In the late 1970s, the National Aeronautics and Space Administration (NASA) initiated development of the SCARE [1,2] (Structural Ceramics Analysis and Reliability Evaluation) computerized design program under the Ceramic Applications in Turbine Engines (CATE) program sponsored by the Department of Energy (DOE) and NASA. The goal of this project was to create public domain ceramics design software to support projects for the DOE, other government agencies, and industry. SCARE has since evolved into the CARES [3–5] (Ceramics Analysis and Reliability Evaluation of Structures) and, most recently, CARES/LIFE [6–9] (Ceramics Analysis and Reliability Evaluation of Structures Life Prediction Program) design programs. CARES is used worldwide by numerous companies and organizations representing various industries such as automotive, aerospace, electronics, and nuclear. CARES/LIFE is currently being evaluated by companies and organizations within the United States.

CARES/LIFE FUNCTIONS

Probabilistic component design involves predicting the probability of failure for a thermomechanically loaded component from specimen rupture data. Typically, these experiments are performed using many simple geometry flexural or tensile test specimens. A static, dynamic, or cyclic load is applied to each specimen until fracture. Statistical strength and SCG (fatigue) parameters are then determined from these data. Using these parameters and the results obtained from a finite element analysis, the time-dependent reliability for a complex component geometry and loading is then predicted. Appropriate design changes are made until an acceptable probability of failure has been reached (Figure 1.2). This design methodology combines the statistical nature of strength-controlling flaws with the mechanics of crack growth to allow for multiaxial stress states, concurrent (simultaneously occurring) flaw populations, and subcritical crack growth. These issues are addressed within the CARES/LIFE program.

CARES/LIFE predicts the probability of failure of a monolithic ceramic component as a function of service time. It assesses the risk that the component will fracture prematurely as a result of subcritical crack growth. The effect of proof testing components prior to service is also considered. CARES/LIFE is coupled to commercially available finite-element programs such as ANSYS, ABAQUS, MSC/NASTRAN, and COSMOS/M. It also retains all of the capabilities of the previous CARES code, which include fast-fracture component reliability evaluation and Weibull parameter estimation from inert strength (without SCG contributing to failure) specimen data. CARES/LIFE can estimate parameters that characterize SCG from specimen data as well.

Finite-element heat transfer and linear-elastic stress analyses are used to determine the component's temperature and stress distributions. The reliability at each element is calculated assuming that randomly distributed volume flaws and/or surface flaws control the failure response. The overall component reliability is the product of all the element survival

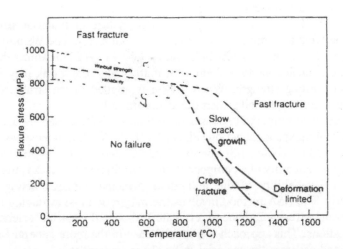

Figure 1.1 Fracture mechanism map for hot-pressed silicon nitride (HPSN) flexure bars [43]. Fracture mechanism maps help illustrate the relative contribution of various failure modes as a function of temperature and stress.

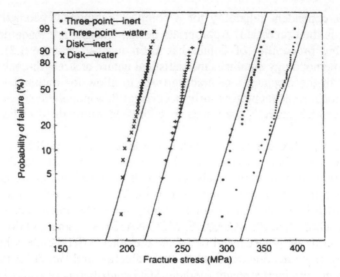

Figure 1.2 Weibull plot of alumina three-point bend bars and biaxially loaded disks fractured in inert and water environments [26]. The solid lines show CARES/LIFE predictions made using Weibull and power law fatigue parameters obtained from four-point bend bar fast-fracture and dynamic fatigue (constant-stress-rate loading) data. Strength degradation in water is predicted for a dynamic load of 1 MPa/s. A mixed-mode fracture criterion was chosen to account for the change in surface flaw reliability for multiaxial stress states.

probabilities. CARES/LIFE generates a data file containing element risk-of-rupture intensities (a local measure of reliability) for graphical rendering of the structure's critical regions.

CARES/LIFE describes the probabilistic nature of material strength, using the Weibull cumulative distribution function [10]. The Weibull equation is based on the weakest-link theory (WLT). WLT assumes that the structure is analogous to a chain with many links. Each link may have a different limiting strength. When a load is applied to the structure such that the weakest link fails, then the structure fails.

The effect of multiaxial stresses on reliability is predicted by using the principle of independent action (PIA) [11,12], the Weibull normal stress averaging method (NSA) [13], or the Batdorf theory [14,15]. For the PIA model, the reliability of a component under multiaxial stresses is the product of the reliability of the individual principal stresses acting independently. The NSA method involves the integration and averaging of tensile normal stress components evaluated about all possible orientations and locations. This approach is a special case of the more general Batdorf theory and assumes the material to be shear insensitive.

The Batdorf theory combines the weakest link theory and linear elastic fracture mechanics (LEFM). Conventional fracture mechanics analysis requies that both the size of the critical crack and its orientation relative to the applied loads determine the fracture stress. The Batdorf theory includes the calculation of the combined probability of the critical flaw being within a certain size range and being located and oriented so that it may cause fracture. A user-selected flaw geometry and a mixed-mode fracture criterion are required to model volume- or surface-strength–limiting defects. Mixed-mode fracture refers to the ability of a crack to grow under the combined actions of a normal load (opening mode) and shear load (sliding and tearing modes) on the crack face. CARES/LIFE includes the total strain energy release rate fracture criterion, which assumes a crack will extend in its own plane (coplanar) [15]. Out-of-plane crack extension criteria are approximated by a simple semiempirical equation [16,17]. Available flaw geometries include the Griffith crack, penny-shaped crack, semicircular crack, and notched crack. The Batdorf theory is equivalent to the probabilistic multiaxial theories proposed by Evans [18] and Matsuo [19].

MODELING SUBCRITICAL CRACK GROWTH

Subcritical crack growth is difficult to model because it is a complex phenomenon often involving a combination of failure mechanisms. Existing models usually involve empirically derived crack propagation laws that describe the crack growth in terms of the stress intensity factor at the crack tip, plus additional parameters obtained from experimental data. The stress intensity factor in LEFM is proportional to the remote applied stress and the crack configuration.

In CARES/LIFE, the relations describing subcritical crack growth are directly incorporated into the PIA, NSA, and Batdorf theories. Subcritical crack growth is modeled with the power law [20,21], the Paris law [22], and the Walker law [23,24] for static and constant-amplitude cyclic loading (Figure 1.3). These laws use experimentally determined parameters that are material- and environment-sensitive. The power law is an exponential relationship between the crack velocity and the stress intensity factor. It is used to model stress corrosion cracking in materials such as glasses and alumina exposed to H_2O. Elevated-temperature, slow crack growth of silicon nitrides, silicon carbides, and alumina also follows power law behavior.

Some polycrystalline ceramics are prone to strength degradation due to mechanical damage induced by cyclic loading. The Paris and Walker laws have been suggested as models to account for this behavior [24]. The Paris law is an exponential relationship between the incremental crack growth per load cycle and the range of the stress intensity factor. The Walker equa-

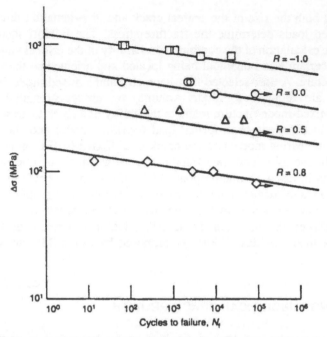

Figure 1.3 Stress amplitude–failure cycles (S-N) plot of 3 mol%-yttria-stabilized zirconia tensile specimens for various R ratios [7,27]. Solid lines show CARES/LIFE predictions at 50% reliability using the Walker SCG law to predict strength degradation due to cyclic fatigue.

tion is functionally similar to the Paris equation, with additional terms to account for the effect of the R-ratio (minimum cycle stress to maximum cycle stress) on lifetime.

CARES/LIFE is capable of predicting the change in a surviving component's reliability after proof testing is performed. Proof testing is the loading of all components prior to service to eliminate those that may fail prematurely. The components that survive the proof test will have a lower (attenuated) risk of failure in service. In CARES/LIFE, the attenuated failure probability is calculated using the PIA, the Weibull normal stress averaging, and the Batdorf theories. The Batdorf model is used to calculate the attenuated failure probability when the proof test load and the service load are not in line or have different multiaxial stress states. This feature is useful when the proof test does not identically simulate the actual service conditions on the component. The durations of the proof test and the service load are also considered in the analysis.

Predicted lifetime reliability of structural ceramic components depends on Weibull and fatigue parameters estimated from widely used tests involving flexural or tensile specimens. CARES/LIFE estimates fatigue parameters from naturally flawed specimens ruptured under static, cyclic, or dynamic (constant stress rate) loading. Fatigue and Weibull parameters are calculated from rupture data of three-point and four-point flexure bars, as well as tensile specimens. For other specimen geometries, a finite element model of the specimen is also required when estimating these parameters.

Activities regarding validation of CARES/LIFE software include example problems obtained from the technical literature, in-house experimental work, Monte Carlo simulations (computer-generated data sets), beta-testing by users, and participation in a round robin study of probabilistic design methodology and corresponding numerical algorithms [25]. Two problems from the open literature are shown, which test the ability of CARES/LIFE to predict the strength of alumina disks and bars under dynamic fatigue [26] and zirconia tensile specimens under cyclic fatigue for various *R*-ratios [27].

ENGINE APPLICATIONS

CARES and CARES/LIFE software is used to design ceramic parts for a wide range of applications. These include parts for turbine and internal combustion engines, bearings, laser windows on test rigs, radomes, radiant heater tubes, spacecraft activation valves and platforms, cathode ray tubes (CRTs), rocket launcher tubes, and ceramic packaging for microprocessors. Engineers and material scientists also use these programs to reduce data from speciment tests to obtain statistical parameters for material characterization. In this way, a comparison of materials is achieved that is independent of the specimen geometry and loading.

The primary thrust behind CARES/LIFE is to support development of advanced heat engines and related ceramics technology infrastructure. DOE and Oak Ridge National Laboratory (ORNL) have several ongoing programs such as the Advanced Turbine Technology Applications Project (ATTAP) [28,29] for automotive gas turbine development, the Heavy Duty Transport Program for low-heat-rejection heavy duty diesel engine development, and the Ceramic Stationary Gas Turbine (CSGT) program for electric power cogeneration. CARES and CARES/LIFE are used in these projects to design stationary and rotating equipment, including turbine rotors, vanes, scrolls, combustors, insulating rings, and seals. These programs are also integrated with the DOE/ORNL Ceramic Technology Project [30] (CTP) characterization and life prediction efforts [31,32].

Solar Turbines Incorporated is using CARES/LIFE to design hot-section turbine parts for the CSGT development program [33] sponsored by the DOE Office of Industrial Technology. This project seeks to replace metallic hot section parts with uncooled ceramic components in an existing design for a natural gas–fired industrial turbine engine operating at a turbine rotor inlet temperature of 1120°C. At least one stage of blades and vanes, as well as the combustor liner, will be replaced with ceramic parts. Ultimately, demonstration of the technology will be proved with a 4000-h engine field test.

Ceramic automotive turbocharger wheels are being developed at Allied-Signal's Turbocharging and Truck Brake Systems [34]. AlliedSignal has designed the CTV7301 silicon nitride turbocharger rotor for the Caterpillar 3406E diesel engine. The reduced rotational inertia of the silicon nitride ceramic compared to a metallic rotor significantly enhances the turbocharger transient performance and reduces emissions. AlliedSignal's effort represents the first design and large-scale deployment of ceramic turbochargers in the United States. Over 1700 units have been supplied to Caterpillar Tractor Company for on-highway truck engines, and these units together have accumulated a total of over 120 million miles of service.

Extensive work has been performed at the Fluid Systems Division of AlliedSignal Aerospace to analyze graphite and ceramic structural components such as high-temperature valves, test fixtures, and turbine wheels. A silicon nitride turbine wheel has been designed as a retrofit for Waspalloy in a military cartridge-mode air turbine starter [35]. The substituted part reduces cost and weight while increasing resistance to temperature, erosion, and corrosion.

Ceramic pistons for a constant-speed drive are being developed at Sundstrand Corporation's Aerospace Division. Constant-speed drives are used to convert variable engine speed to a constant output speed for aircraft electrical generators. The calculated probability of failure of the piston is less than 0.2×10^{-8} under the most severe limit-load condition. This program is sponsored by the U.S. Navy and ARPA (Advanced Research Projects Agency, formerly DARPA—Defense Advanced Research Projects Agency). Sundstrand has designed ceramic components for a number of other applications, most notably for aircraft auxiliary power units.

GTE Laboratories has analyzed and designed a ceramic-to-metal brazed joint for automotive gas turbine engines [36,37]. A major problem in ceramic-to-metal joining is the thermal expansion mismatch between the two different materials. This results in high residual stresses that increase the likelihood of ceramic failure. One of the goals of this work was to improve the capability of the metal shaft to transmit power by reducing concentrated tensile stresses. Their results confirmed the importance of probabilistic failure analysis for assessing the performance of various brazed joint designs.

A monolithic graphite spacecraft activation valve was designed by the Aerospace and Electronics Division of Boeing Space Defense Group [38].

The valve directs reaction control gases for fine-tuning a high-performance kinetic energy kill vehicle's trajectory during the last 9 s of flight. The valve was designed to withstand a gas pressure of 11.4 MPa at 1930°C.

At the NASA-Lewis Research Center, a design study demonstrated the viability of an uncooled silicon nitride combustor for commercial application in a 300-kW engine with a turbine inlet temperature of 1370°C [39]. The analysis was performed for the worst transient thermal stress in an emergency shutdown. The most critical area was found to be around the dilution port.

Ceramic poppet valves for spark ignition engines have been designed by TRW's Automotive Valve Division [40], as well as by General Motors. These parts have been field tested in passenger cars with excellent results. Potential advantages offered by these valves include reduced seat insert and valve guide wear, improved valve train dynamics, increased engine output, and reduced friction loss using lower spring loads.

DESIGNING OPTICAL AND OTHER COMPONENTS

The largest known zinc-selenide (ZnSe) window has been designed by Hughes Danbury Optical Systems (formerly Perkin-Elmer). The window formed a pressure barrier between a cryogenic vacuum chamber containing optical equipment and a sensor chamber. The window measured 79 cm in diameter, was 2.5 cm thick, and was used in a test facility at Boeing for long-range infrared sensors.

Glass components behave in a similar manner as ceramics and must be designed using reliability evaluation techniques. Phillips Display Components Company has analyzed the possibility of alkali strontium silicate glass CRTs spontaneously imploding [41]. CRTs are under a constant static load due to the pressure forces placed on the outside of the evacuated tube. A 68-cm diagonal tube was analyzed with and without an implosion protection band. The implosion protection band reduces the overall stresses in the tube and, in the event of an implosion, also contains the glass particles within the enclosure. Stress analysis showed compressive stresses on the front face and tensile stresses on the sides of the tube. The implosion band reduced the maximum principal stress by 20%. Reliability analysis showed that the implosion protection band significantly reduced the probability of failure to about 5×10^{-5}.

The structural integrity of a silicon carbide convective air-heater for use in an advanced power generation system has been assessed by ORNL and the NASA-Lewis Research Center. The design used a finned tube arrangement 1.8 m in length with 2.5-cm diameter tubes. Incoming air was to be heated from 390°C to 700°C. The hot gas flow across the tubes was at

980°C. Heat transfer and stress analyses revealed that maximum stress gradients across the tube wall nearest the incoming air would be the most likely source of failure.

At the University of Florida College of Dentistry, probabilistic design techniques are being applied to dental ceramic crowns. Frequent failure of some ceramic crowns (e.g., 35% failure of molar crowns after three years), which occurs because of residual and functional stresses, necessi-

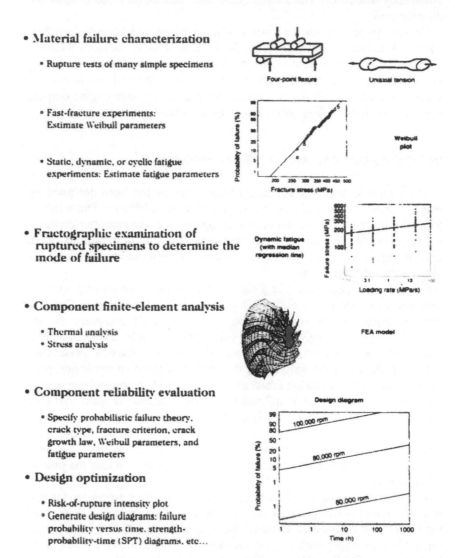

- Material failure characterization

    - Rupture tests of many simple specimens

        Four-point flexure   Uniaxial tension

    - Fast-fracture experiments: Estimate Weibull parameters

        Weibull plot

    - Static, dynamic, or cyclic fatigue experiments: Estimate fatigue parameters

- Fractographic examination of ruptured specimens to determine the mode of failure

    Dynamic fatigue (with median regression line)

- Component finite-element analysis

    - Thermal analysis
    - Stress analysis

        FEA model

- Component reliability evaluation

    - Specify probabilistic failure theory, crack type, fracture criterion, crack growth law, Weibull parameters, and fatigue parameters

- Design optimization

    - Risk-of-rupture intensity plot
    - Generate design diagrams: failure probability versus time, strength-probability-time (SPT) diagrams, etc...

Figure 1.4 Probabilistic component design procedure.

tates design modifications and improvement of these restorations. The University of Florida is investigating thermal tempering treatment as a means of introducing compressive stresses on the surface of dental ceramics to improve the resistance to failure [42]. Evaluation of the risk of material failure must be considered not only for the service environment but also from the tempering process (Figure 1.4).

FUTURE DIRECTIONS

The NASA-developed CARES software has been successfully used to design ceramic components for many demanding applications. The latest version of this program, CARES/LIFE, allows the engineer to design against premature failure from progressive cracking due to subcritical crack growth. CARES/LIFE is a new program and has been selectively distributed. Potential enhancements include transient creep analysis, three-parameter Weibull statistics, oxidation modeling, flaw anisotropy, threshold stress behavior, and crack-resistance curve modeling. When officially released, CARES/LIFE will be distributed through the NASA Computer Software Management and Information Center (COSMIC), at the University of Georgia.

ACKNOWLEDGEMENTS

The authors want to thank all the CARES and CARES/LIFE users referenced in this article. Further acknowledgement is given to those who provided text and figures, including D. Baker, A. Ghosh, S. Gupta, B. Harkins, B. Hojjatie, and B. Skrobacz.

REFERENCES

1 Gyekenyesi, J. P. "SCARE: A Postprocessor Program to MSC/NASTRAN for Reliability Analysis of Structural Ceramic Components," *Journal of Engineering for Gas Turbines and Power*, Vol. 108, 1986, pp. 540–546.

2 Gyekenyesi, J. P. and Nemeth, N. N. "Surface Flaw Reliability Analysis of Ceramic Components with the SCARE Finite Element Postprocessor Program," *Journal of Engineering for Gas Turbines and Power*, Vol. 109, 1987, pp. 274–281.

3 Nemeth, N. N., Manderscheid, J. M., and Gyekenyesi, J. P. "Designing Ceramic Components with the CARES Computer Program," *American Ceramic Society Bulletin*, Vol. 68, 1989, pp. 2064–2072.

4 Nemeth, N. N., Manderscheid, J. M., and Gyekenyesi, J. P. "Ceramic Analysis and Reliability Evaluation of Structures (CARES)," NASA TP-2916, Aug. 1990.

5 Powers, L. M., Starlinger, A., and Gyekenyesi, J. P. "Ceramic Component Reliability with the Restructured NASA/CARES Computer Program," NASA TM-105856, Sept. 1992.

6 Nemeth, N. N., Powers, L. M., Janosik, L. A., and Gyekenyesi, J. P. "Time-Dependent Reliability Analysis of Monolithic Ceramic Components Using the CARES/LIFE Integrated Design Program," *Life Prediction Methodologies and Data for Ceramic Materials, ASTM STP 1201*, C. R. Brinkman, and S. F. Duffy, eds., American Society for Testing and Materials, Philadelphia, 1993, pp. 390–408.

7 Nemeth, N. N., Powers, L. M., Janosik, L. A., and Gyekenyesi, J. P. "Lifetime Reliability Evaluation of Structural Ceramic Parts with the CARES/LIFE Computer Program," AIAA paper 93-1497-CP, *Proceedings of the 34th AIAA/ASME/ASCE/ASC Structures, Structural Dynamics, and Materials Conference*, April 19–21, 1993, La Jolla, California, American Institute for Aeronautics and Astronautics, Washington, D.C., 1993, pp. 1634–1646.

8 Powers, L. M., Janosik, L. A., Nemeth, N. N., and Gyekenyesi, J. P. "Lifetime Reliability Evaluation of Monolithic Ceramic Components Using the CARES/LIFE Integrated Design Program," *Proceedings of the American Ceramic Society Meeting and Exposition*, Cincinnati, Ohio, April 19–22, 1993.

9 Nemeth, N. N., Powers, L. M., Janosik, L. A., and Gyekenyesi, J. P. "Ceramics Analysis and Reliability Evaluation of Structures Life Prediction Program (CARES/LIFE) Users and Programmers Manual," NASA TM-106316, to be published.

10 Weibull, W. A. "A Statistical Theory of the Strength of Materials," *Ingenoirs Vetenskaps Akadanien Handlinger*, 1939, No. 151.

11 Barnett, R. L., Connors, C. L., Hermann, P. C., and Wingfield, J. R. "Fracture of Brittle Materials under Transient Mechanical and Thermal Loading," U.S. Air Force Flight Dynamics Laboratory, AFFDL-TR-66-220, 1967 (NTIS AD-649978).

12 Freudenthal, A. M. "Statistical Approach to Brittle Fracture," *Fracture, Vol. 2: An Advanced Treatise, Mathematical Fundamentals*, H. Liebowitz, ed., Academic Press, 1968, pp. 591–619.

13 Weibull, W. A. "The Phenomenon of Rupture in Solids," *Ingenoirs Vetenskaps Akadanien Handlinger*, 1939, No. 153.

14 Batdorf, S. B. and Crose, J. G. "A Statistical Theory for the Fracture of Brittle Structures Subjected to Nonuniform Polyaxial Stresses," *Journal of Applied Mechanics*, Vol. 41, No. 2, 1974, pp. 459–464.

15 Batdorf, S. B. and Heinisch, H. L., Jr. "Weakest Link Theory Reformulated for Arbitrary Fracture Criterion," *Journal of the American Ceramic Society*, Vol. 61, No. 7–8, 1978, pp. 355–358.

16 Palaniswamy, K. and Knauss, W. G. "On the Problem of Crack Extension in Brittle Solids under General Loading," *Mechanics Today*, Vol. 4, 1978, pp. 87–148.

17 Shetty, D. K. "Mixed-Mode Fracture Criteria for Reliability Analysis and Design with Structural Ceramics," *Journal of Engineering for Gas Turbines and Power*, Vol. 109, No. 3, 1987, pp. 282–289.

18 Evans, A. G. "A General Approach for the Statistical Analysis of Multiaxial Fracture," *Journal of the American Ceramic Society*, Vol. 61, 1978, pp. 302–306.

19 Matsuo, Y. *Trans. of the Japan Society of Mechanical Engineers*, Vol. 46, 1980, pp. 605–611.

20 Evans, A. G. and Wiederhorn, S. M. "Crack Propagation and Failure Prediction in Silicon Nitride at Elevated Temperatures," *Journal of Material Science*, Vol. 9, 1974, pp. 270–278.

21 Wiederhorn, S. M. *Fracture Mechanics of Ceramics*. R. C. Bradt, D. P. Hasselman, and F. F. Lange, eds., Plenum, New York, 1974, pp. 613–646.

22 Paris, P. and Erdogan, F. "A Critical Analysis of Crack Propagation Laws," *Journal of Basic Engineering*, Vol. 85, 1963, pp. 528–534.

23 Walker, K. "The Effect of Stress Ratio during Crack Propagation and Fatigue for 2024-T3 and 7075-T6 Aluminum," *ASTM STP 462, Effects of Environment and Complex Load History on Fatigue Life*, 1970, pp. 1–14.

24 Dauskardt, R. H., James, M. R., Porter, J. R. and Ritchie, R. O. "Cyclic Fatigue Crack Growth in SiC-Whisker-Reinforced Alumina Ceramic Composite: Long and Small Crack Behavior," *Journal of the American Ceramic Society*, Vol. 75, No. 4, 1992, pp. 759–771.

25 Powers, L. M. and Janosik, L. A. "A Numerical Round Robin for the Reliability Prediction of Structural Ceramics," AIAA paper 93-1498-CP, *Proceedings of the 34th AIAA/ASME/ASCE/ASC Structures, Structural Dynamics, and Materials Conference*, April 19–21, 1993, La Jolla, California, pp. 1647–1658.

26 Chao, L. Y. and Shetty, D. K. "Time-Dependent Strength Degradation and Reliability of an Alumina Ceramic Subjected to Biaxial Flexure," *Life Prediction Methodologies and Data for Ceramic Materials, ASTM STP 1201*, C. R. Brinkman and S. F. Duffy, eds., American Society for Testing and Materials, Philadelphia, PA, 1993, pp. 228–249.

27 Liu, S. Y. and Chen, I.-W. "Fatigue of Yttria-Stabilized Zirconia: I, Fatigue Damage, Fracture Origins, and Lifetime Prediction," *Journal of the American Ceramic Society*, Vol. 74, No. 6, 1991, pp. 1197–1205.

28 Smyth, J. R., Morey, R. E., and Schultz, R. W. "Ceramic Gas Turbine Technology Development and Applications," ASME Paper 93-GT-361, Presented at the *International Gas Turbine and Aeroengine Congress and Exposition*, Cincinnati, Ohio, May 24–27, 1993.

29 Berenyi, S. G., Hilpisch, S. J., and Groseclose, L. E. "Advanced Turbine Technology Applications Project (ATTAP)," *Proceedings of the Annual Automotive Technology Development Contractor's Coordination Meeting*, P-278, Dearborn, Michigan, October 18–21, 1993, SAE International, Warrendale, PA, pp. 249–255.

30 Johnson, D. R. and Schultz, R. B. "The Ceramic Technology Project: Ten Years of Progress," ASME Paper 93-GT-417, Presented at the *International Gas Turbine and Aeroengine Congress and Exposition*, Cincinnati, Ohio, May 24–27, 1993.

31 Cuccio, J., Peralta, A. D., Wu, D. C., Fang, H. T., Song, Z. S., Brehm, P. J., Menon, N. M., and Strangman, T. E. "Life Prediction Methodology for Ceramic Components of Advanced Heat Engines," *Proceedings of the Annual Automotive Technology Development Contractor's Coordination Meeting*, P-278, Dearborn, Michigan, October 18–21, 1993, pp. 185–198.

32 Khandelwal, P. K., Provenzano, N. J., and Schneider, W. E. "Life Prediction Methodology for Ceramic Components of Advanced Vehicular Engines," *Proceedings of the Annual Automotive Technology Development Contractor's Coordination Meeting*, P-278, Dearborn, Michigan, October 18–21, 1993, pp. 199–211.

33 van Roode, M., Brentnall, W. D., Norton, P. F., and Pytanowski, G. P. "Ceramic Stationary Gas Turbine Development," ASME Paper 93-GT-309, Presented at the

International Gas Turbine and Aeroengine Congress and Exposition, Cincinnati, Ohio, May 24–27, 1993.

34 Baker, C. and Baker, D. "Design Practices for Structural Ceramics in Automotive Turbocharger Wheels," *Engineered Materials Handbook, Volume 4: Ceramics and Glasses*, ASM International, 1991, pp. 722–727.

35 Poplawsky, C. J., Lindberg, L., Robb, S., and Roundy, J. "Development of an Advanced Ceramic Turbine Wheel for an Air Turbine Starter," SAE Paper 921945, Presented at *Aerotech '92*, Anaheim, California, October 5–8, 1992.

36 Selverian, J. H., O'Neil, D., and Kang, S. "Ceramic-to-Metal Joints: Part I – Joint Design," *American Ceramic Society Bulletin*, Vol. 71, No. 9, 1992, pp. 1403–1409.

37 Selverian, J. H. and Kang, S. "Ceramic-to-Metal Joints: Part II – Performance and Strength Prediction," *American Ceramic Society Bulletin*, Vol. 71, No. 10, 1992, pp. 1511–1520.

38 Snydar, C. L. "Reliability Analysis of a Monolithic Graphite Valve," Presented at the *15th Annual Conference on Composites, Materials, and Structures*, Cocoa Beach, FL, 1991.

39 Salem, J. A., Manderscheid, J. M., Freedman, M. R., and Gyekenyesi, J. P. "Reliability Analysis of a Structural Ceramic Combustion Chamber," ASME Paper 91-GT-155, Presented at the *International Gas Turbine and Aeroengine Congress and Exposition*, Orlando, Florida, June 3–6, 1991.

40 Wills, R. R. and Southam, R. E. "Ceramic Engine Valves," *Journal of the American Ceramic Society*, Vol. 72, No. 7, 1989, pp. 1261–1264.

41 Ghosh, A., Cha, C. Y., Bozek, W., and Vaidyanathan, S. "Structural Reliability Analysis of CRTs," *Society for Information Display International Symposium Digest of Technical Papers Volume XXIII*, Hynes Convention Center, Boston, Massachusetts, May 17–22, 1992, Society of Information Display, Playa Del Ray, CA, pp. 508–510.

42 Hojjatie, B. "Thermal Tempering of Dental Ceramics," *Proceedings of the ANSYS Conference and Exhibition, Vol. 1*, Swanson Analysis Systems Inc., Houston, PA, 1992, pp. I.73–I.91.

43 Quinn, G. D. "Fracture Mechanism Maps for Advanced Structural Ceramics: Part 1; Methodology and Hot-Pressed Silicon Nitride Results," *Journal of Materials Science*, Vol. 25, 1990, pp. 4361–4376.

CHAPTER 2

Durability Evaluation of Ceramic Components Using CARES/LIFE

NOEL N. NEMETH,[1] LYNN M. POWERS,[2]
LESLEY A. JANOSIK[1] and JOHN P. GYEKENYESI[1]

OVERVIEW

THE computer program CARES/LIFE calculates the time-dependent reliability of monolithic ceramic components subjected to thermo-mechanical and/or proof test loading. This program is an extension of the CARES (Ceramics Analysis and Reliability Evaluation of Structures) computer program. CARES/LIFE accounts for the phenomenon of subcritical crack growth (SCG) by utilizing the power law, Paris law, or Walker equation. The two-parameter Weibull cumulative distribution function is used to characterize the variation in component strength. The effects of multiaxial stresses are modeled using either the principle of independent action (PIA), the Weibull normal stress averaging method (NSA), or the Batdorf theory. Inert strength and fatigue parameters are estimated from rupture strength data of naturally flawed specimens loaded in static, dynamic, or cyclic fatigue. Application of this design methodology is demonstrated using experimental data from alumina bar and disk flexure specimens, which exhibit SCG when exposed to water.

NOMENCLATURE

A surface area; material-environmental fatigue constant
a crack half length
B subcritical crack growth constant
$\bar{c}$ Shetty's constant in mixed-mode fracture criterion
g g-factor

[1]NASA Lewis Research Center, Cleveland, OH, U.S.A.
[2]Department of Civil Engineering, Cleveland State University, Cleveland, OH, U.S.A.

H step function
i ranking of ordered fracture data in statistical analysis
K stress intensity factor
k crack density coefficient
m Weibull modulus, or shape parameter
N material-environmental fatigue constant
n number of cycles
P_f cumulative failure probability
Q cyclic fatigue parameter
R ratio of minimum to maximum effective stress in a load cycle
T period of one cycle
t time
t_0 time-dependent scale parameter
V volume; crack velocity
x,y,z Cartesian coordinate directions
Y crack geometry factor

GREEK LETTERS

α angle between σ_n and the stress σ_1
β angle between σ_n projection and the stress σ_2 in plane perpendicular to σ_1
Δ increment
η crack density function
π 3.1416
σ applied stress distribution
σ_0 Weibull scale parameter
$\sigma_1, \sigma_2, \sigma_3$ tensor stress components; principal stresses ($\sigma_1 \geq \sigma_2 \geq \sigma_3$)
τ shear stress acting on oblique plane, whose normal is determined by angles α and β
Ψ spatial location (x,y,z) and orientation (α,β) in a component
Ω solid angle in three-dimensional principal stress space for which $\sigma_e \geq \sigma_{cr}$
ω angle in two-dimensional principal stress space for which $\sigma_e \geq \sigma_{cr}$

SUBSCRIPTS

B Batdorf
c cyclic; critical
ch characteristic
cr critical
d dynamic fatigue

e,ef effective
eq equivalent
  f failure; fracture
  I crack opening mode
  II crack sliding mode
  III crack tearing mode
  i ith
max maximum
min minimum
  n normal; normal stress averaging
  S surface
  T transformed
  u uniaxial
  V volume
  w Weibull
  θ characteristic

SUPERSCRIPTS

  ~ modified parameter
  — normalized quantity

INTRODUCTION

Advanced ceramic components designed for gasoline, diesel, and turbine heat engines are leading to lower engine emissions, higher fuel efficiency, and more compact designs due to their low density and ability to retain strength at high temperatures. Ceramic materials are also used for wear parts (nozzles, valves, seals, etc.), cutting tools, grinding wheels, bearings, coatings, electronics, and human prostheses. Among the many requirements for the successful application of advanced ceramics are the proper characterization of material properties and the use of a mature and validated brittle material design methodology.

Ceramics are brittle, and the lack of ductility leads to low strain tolerance, low fracture toughness, and large variations in observed fracture strength. The material as processed has numerous inherent randomly distributed flaws. The observed scatter in fracture strength is caused by the variable severity of these flaws. The ability of a ceramic component to sustain a load also degrades over time. This is due to a variety of effects such as oxidation, creep, stress corrosion, and cyclic fatigue. Stress corrosion and cyclic fatigue result in a phenomenon called subcritical crack growth (SCG). SCG initiates at an existing flaw and continues until a critical length is reached, causing catastrophic crack propagation. The SCG failure mechanism is a load-induced phenomenon over time. It can also be a function of chemical reaction, environment, debris wedging near the crack tip, and deterioration of bridging ligaments. Because of this complexity, models that have been developed tend to be semiempirical and approximate the behavior of subcritical crack growth phenomenologically.

The objective of this chapter is to present a description of the integrated design computer program, CARES/LIFE [1] (Ceramics Analysis and Reliability Evaluation of Structures Life Prediction Program). The theory and concepts presented in this chapter reflect the capabilities of the CARES/LIFE program for time-dependent probabilistic design. To determine the validity of the design methodology utilized in this software, time-dependent reliability predictions from CARES/LIFE are compared with experimental data [2] from uniaxially and biaxially loaded alumina flexure bars and disks, which are known to exhibit slow crack growth in water.

PROGRAM CAPABILITY AND DESCRIPTION

Probabilistic component design involves predicting the probability of failure for a thermomechanically loaded component from specimen rupture data. Typically, these experiments are performed using many simple geometry flexural or tensile test specimens. A static, dynamic, or cyclic load is applied to each specimen until fracture. Statistical strength and SCG (fatigue) parameters are then determined from these data. Using these parameters and the results obtained from a finite element analysis, the time-dependent reliability for a complex component geometry and loading is then predicted. Appropriate design changes are made until an acceptable probability of failure has been reached. This design methodology combines the statistical nature of strength-controlling flaws with the mechanics of crack growth to allow for multiaxial stress states, concurrent (simultaneously occurring) flaw populations, and scaling effects. These issues are addressed within the CARES/LIFE program.

CARES/LIFE predicts the probability of failure of a monolithic ceramic component as a function of service time. It assesses the risk that the component will fracture prematurely as a result of subcritical crack growth. The effect of proof testing components prior to service is also considered. CARES/LIFE is coupled to commercially available finite-element programs such as ANSYS, ABAQUS, MSC/NASTRAN, and COSMOS/M. CARES/LIFE is an extension of the CARES [3,4] program. It retains all of the capabilities of the previous CARES code, which include fast-fracture component reliability evaluation and Weibull parameter estimation from inert strength (without SCG contributing to failure) specimen data. CARES/LIFE can estimate parameters that characterize SCG from specimen data as well.

Finite-element heat transfer and linear-elastic stress analyses are used to determine the component's temperature and stress distributions. The reliability at each element is calculated, assuming that randomly distributed volume flaws and/or surface flaws control the failure response. The probability of survival for each element is assumed to be a mutually exclusive event, and the overall component reliability is then the product of all the element survival probabilities. CARES/LIFE generates a data file containing element risk-of-rupture intensities (a local measure of reliability) for graphical rendering of the structure's critical regions.

CARES/LIFE describes the probabilistic nature of material strength, using the Weibull cumulative distribution function. The effect of multiaxial stresses on reliability is predicted using the principle of independent action (PIA) [5,6], the Weibull normal stress averaging method (NSA) [7], or the Batdorf theory [8,9]. The Batdorf theory combines the weakest link theory and linear elastic fracture mechanics (LEFM). Conventional fracture mechanics analysis requires that both the size of the critical crack and its orientation relative to the applied loads determine the fracture stress. The Batdorf theory includes the calculation of the combined probability of the critical flaw being within a certain size range and being located and oriented so that it may cause fracture. A user-selected flaw geometry and a mixed-mode fracture criterion are required to model volume- or surface-strength–limiting defects. Mixed-mode fracture refers to the ability of a crack to grow under the combined actions of a normal load (opening mode) and shear load (sliding and tearing modes) on the crack face. CARES/LIFE includes the total strain energy release rate fracture criterion, which assumes a crack will extend in its own plane (coplanar) [9]. Out-of-plane crack extension criteria are approximated by a simple semi-empirical equation [10,11]. Available flaw geometries include the Griffith crack, penny-shaped crack, semicircular crack, and notched crack. The Batdorf theory is equivalent to the probabilistic multiaxial theories proposed by Evans [12] and Matsuo [13].

Subcritical crack growth is difficult to model, because it is a complex phenomenon often involving a combination of failure mechanisms. Existing models usually involve empirically derived crack propagation laws that describe the crack growth in terms of the stress intensity factor at the crack tip, plus additional parameters obtained from experimental data.

In CARES/LIFE, the relations describing subcritical crack growth are directly incorporated into the PIA, NSA, and Batdorf theories. Subcritical crack growth is modeled with the power law [14,15], the Paris law [16], and the Walker law [17,18] for static and constant-amplitude cyclic loading. These laws use experimentally determined parameters that are material-

and environment-sensitive. The power law is used to model stress corrosion cracking in materials such as glasses and alumina exposed to H_2O. Elevated-temperature slow crack growth of silicon nitrides, silicon carbides, and alumina also follows power law behavior.

Some polycrystalline ceramics are prone to strength degradation due to mechanical damage induced by cyclic loading. The Paris and Walker laws have been suggested as models to account for this behavior [18]. The Walker equation is functionally similar to the Paris equation with additional terms to account for the effect of the R-ratio (minimum cycle stress to maximum cycle stress) on lifetime.

CARES/LIFE is capable of predicting the change in a surviving component's reliability after proof testing is performed. Proof testing is the loading of all components prior to service to eliminate those that may fail prematurely. The components that survive the proof test will have a lower (attenuated) risk of failure in service. In CARES/LIFE, the attenuated failure probability is calculated using the PIA, the Weibull normal stress averaging, and the Batdorf theories. The Batdorf model is used to calculate the attenuated failure probability when the proof test load and the service load are not in line or have different multiaxial stress states. This feature is useful when the proof test does not identically simulate the actual service conditions on the component. The durations of the proof test and the service load are also considered in the analysis.

Predicted lifetime reliability of structural ceramic components depends on Weibull and fatigue parameters estimated from widely used tests involving flexural or tensile specimens. CARES/LIFE estimates fatigue parameters from naturally flawed specimens ruptured under static, cyclic, or dynamic (constant stress rate) loading. Fatigue and Weibull parameters are calculated from rupture data of three-point and four-point flexure bars, as well as tensile specimens. For other specimen geometries, a finite element model of the specimen is also required when estimating these parameters.

THEORY

Time-dependent reliability is based on the mode I equivalent stress distribution transformed to its equivalent stress distribution at time $t = 0$. Investigations of mode I crack extension [19] have resulted in the following relationship for the equivalent mode I stress intensity factor:

$$K_{Ieq}(\Psi,t) = \sigma_{Ieq}(\Psi,t)Y\sqrt{a(\Psi,t)} \tag{1}$$

where $\sigma_{Ieq}(\Psi,t)$ is the equivalent mode I stress on the crack, Y is a function of crack geometry, $a(\Psi,t)$ is the appropriate crack length, and Ψ represents a location (x,y,z) within the body and the orientation (α,β) of the crack. In some models such as the Weibull and PIA, Ψ represents a location only. Y is a function of crack geometry; however, herein, it is assumed constant with subcritical crack growth. Crack growth as a function of the equivalent mode I stress intensity factor is assumed to follow a power law relationship:

$$\frac{da(\Psi,t)}{dt} = AK_{Ieq}^{N}(\Psi,t) \tag{2}$$

where A and N are material/environmental constants. The transformation of the equivalent stress distribution at the time of failure, $t = t_f$, to its critical effective stress distribution at time $t = 0$ is expressed as [20,21]

$$\sigma_{Ieq,0}(\Psi,t_f) = \left[\frac{\displaystyle\int_0^{t_f} \sigma_{Ieq}^{N}(\Psi,t)dt}{B} + \sigma_{Ieq}^{N-2}(\Psi,t_f) \right]^{1/(N-2)} \tag{3}$$

where

$$B = \frac{2}{AY^2 K_{Ic}^{N-2}(N-2)}$$

is a material/environmental fatigue parameter, K_{Ic} is the critical stress intensity factor, and $\sigma_{Ieq}(\Psi,t_f)$ is the equivalent stress distribution in the component at time $t = t_f$. The dimensionless fatigue parameter N is independent of fracture criterion. B is adjusted to satisfy the requirement that, for a uniaxial stress state, all models produce the same probability of failure. The parameter B has units of stress^2 × time.

VOLUME FLAW ANALYSIS

The probability of failure for a ceramic component using the Batdorf model [8,9,22] for volume flaws is

$$P_{fV} = 1 - \exp\left\{ -\int_V \left[\int_0^{\sigma_{e_{max}}} \frac{\Omega}{4\pi} \frac{d\eta_V(\sigma_{cr})}{d\sigma_{cr}} d\sigma_{cr} \right] dV \right\} \tag{4}$$

where V is the volume, η_V is the crack density function, $\sigma_{e_{max}}$ is the maximum value of $\sigma_{Ieq,0}$ for all values of Ψ, and Ω is the area of a solid angle projected onto a unit radius sphere in principal stress space containing all crack orientations for which the effective stress is greater than or equal to the critical mode I strength, σ_{cr}. The crack density distribution is a function of the critical effective stress distribution. For volume flaw analysis, the crack density function is expressed as

$$\eta_V(\sigma_{cr}(\Psi)) = k_{BV}\sigma_{cr}^{m_V} \tag{5}$$

where k_{BV} and m_V are material constants. The solid angle is expressed as

$$\Omega = \int_0^{2\pi} \int_0^{\pi} H(\sigma_{leq,0},\sigma_{cr}) \sin \alpha d\alpha d\beta \tag{6}$$

where

$$H(\sigma_{leq,0},\sigma_{cr}) = \begin{cases} 1 & \sigma_{leq,0} \geq \sigma_{cr} \\ 0 & \sigma_{leq,0} < \sigma_{cr} \end{cases}$$

and α and β are the radial and azimuthal angles, respectively, on the unit radius sphere. The transformed equivalent stress $\sigma_{leq,0}$ is dependent on the appropriate fracture criterion, crack shape, and time to failure, t_f. Equation (4) can be simplified by performing the integration of σ_{cr} [22], yielding the time-dependent probability of failure for volume flaw analysis:

$$P_{fV}(t_f) = 1 - \exp\left[-\frac{k_{BV}}{2\pi} \int_V \int_0^{2\pi} \int_0^{\pi/2} \sigma_{leq,0}^{m_V}(\Psi,t_f) \sin \alpha d\alpha d\beta dV \right] \tag{7}$$

Fracture criteria and crack shapes available for time-dependent analysis are identical to those used for fast-fracture analysis in CARES [3,4]. These fracture criteria include Weibull normal stress averaging (a shear-insensitive case of the Batdorf theory), the total coplanar strain energy release rate, and the noncoplanar crack extension (Shetty) criterion.

For a stressed component, the probability of failure for volume flaw analysis is calculated from Equation (7). The finite-element method enables discretization of the component into incremental volume elements. CARES/LIFE evaluates the reliability at the Gaussian integration points of the element or, optionally, at the element centroid. Subelement volume is defined as the contribution of the integration point to the element volume in the course of the numerical integration procedure. The volume of each subelement (corresponding to a Gauss integration point) is calculated using shape functions inherent to the element type [4]. Assuming that the probability of survival for each element is a mutually exclusive event, the overall component reliability is then the product of all the calculated element (or subelement) survival probabilities.

SURFACE FLAW ANALYSIS

The probability of failure for a ceramic component using the Batdorf model [8,9,22] for surface flaws is

$$P_{fs} = 1 - \exp\left\{ -\int_A \left[\int_0^{\sigma_{e_{max}}} \frac{\omega}{\pi} \frac{d\eta_s(\sigma_{cr})}{d\sigma_{cr}} d\sigma_{cr} \right] dA \right\} \tag{8}$$

where A is the surface area, η_s is the crack density function, $\sigma_{e_{max}}$ is the maximum value of $\sigma_{Ieq,0}$ for all values of Ψ, and ω is the arc length of an angle α projected onto a unit radius semicircle in principal stress space containing all of the crack orientations for which the effective stress is greater than or equal to the critical stress. Analogous to the argument for volume flaws, Equation (8) can be reformulated, yielding [22]

$$P_{fs}(t_f) = 1 - \exp\left[-\frac{k_{BS}}{\pi} \int_A \int_0^{\tau} \sigma_{Ieq,0}^{m_s}(\Psi, t_f) d\alpha dA \right] \tag{9}$$

The transformed equivalent stress $\sigma_{Ieq,0}$ is dependent on the appropriate fracture criterion, crack shape, and time to failure, t_f. The fracture criteria and crack shapes available for time-dependent analysis are identical to those used for fast-fracture analysis. These fracture criteria include Weibull normal stress averaging (a shear-insensitive case of the Batdorf theory), the total coplanar strain energy release rate, and the noncoplanar crack extension (Shetty) criterion.

The finite-element method enables discretization of the surface of the component into incremental area elements. CARES/LIFE evaluates the failure probability at the Gaussian integration points of shell elements or, optionally, at the element centroids. The area of each subelement (corresponding to a Gaussian integration point) is calculated using shape functions inherent to the element type [4]. Assuming that the probability of survival for each element is a mutually exclusive event, the overall component reliability is then the product of all the calculated element (or subelement) survival probabilities.

STATIC FATIGUE

Static fatigue is defined as the application of a nonvarying load over time. For this case, the mode I equivalent stress, $\sigma_{\text{Ieq}}(\Psi,t)$, is independent of time and is thus denoted by $\sigma_{\text{Ieq}}(\Psi)$. Integrating Equation (3) with respect to time yields

$$\sigma_{\text{Ieq},0}(\Psi,t_f) = \sigma_{\text{Ieq}}(\Psi)\left[\frac{t_f\sigma_{\text{Ieq}}^2(\Psi)}{B} + 1\right]^{1/N-2} \tag{10}$$

DYNAMIC FATIGUE

Dynamic fatigue is defined as the application of a constant stress rate $\dot{\sigma}(\Psi)$ over a period of time, t. Assuming the applied stress is zero at time $t = 0$, then

$$\sigma_{\text{Ieq}}(\Psi,t) = \dot{\sigma}(\Psi)t \tag{11}$$

Substituting Equation (11) into Equation (3) results in an expression for effective stress at the time of failure

$$\sigma_{\text{Ieq},0}(\Psi) = \left[\frac{\sigma_{\text{Ieq}}^N(\Psi,t_f)t_f}{(N + 1)B} + \sigma_{\text{Ieq}}^{N-2}(\Psi,t_f)\right]^{1/N-2} \tag{12}$$

CYCLIC FATIGUE

Cyclic fatigue is defined as the repeated application of a loading sequence. Analysis of the time-dependent probability of failure for a component subjected to various cyclic boundary load conditions is simplified by transforming that type of loading to an equivalent static load. The conversion satisfies the requirement that both systems will cause the same crack growth [23]. Implicit in this conversion is the validity of Equation (2) for describing the crack growth. The probability of failure is obtained with respect to the equivalent static state.

Evans [24] and Mencik [23] defined g-factors, $g(\Psi)$, for various types of cyclic loading, that are used to convert the cyclic load pattern to an equivalent static load. For periodic loading, T is the time interval of one cycle, and $\sigma_{\text{Ieq}}(\Psi)$ is the equivalent static stress acting over the same time interval, t_1, as the applied cyclic stress, $\sigma_{Ieqc}(\Psi,t)$, at some location Ψ. The equivalent static stress is related to the cyclic stress by

$$\sigma_{Ieq}^N(\Psi)t_1 = \int_0^{t_1} \sigma_{Ieqc}^N(\Psi,t)dt = t_1 \left[\frac{\int_0^T \sigma_{Ieqc}^N(\Psi,t)dt}{T} \right] = g(\Psi)\sigma_{Ieqcmax}^N(\Psi)t_1 \quad (13)$$

The CARES/LIFE program uses the maximum cyclic stress, $\sigma_{Ieqcmax}(\Psi)$, of the periodic load as a characteristic value to normalize the g-factor. For a periodic load over a time t_1, the mode I static equivalent stress distribution is

$$\sigma_{Ieq,0}(\Psi,t_f) = \sigma_{Ieqcmax}(\Psi) \left[\frac{g(\Psi)t_f\sigma_{Ieqcmax}^2(\Psi)}{B} + 1 \right]^{1/N-2} \quad (14)$$

The use of g-factors for determining component life is an unconservative practice for materials prone to cyclic damage. The Walker equation [17], which has traditionally been used in metals design, has been suggested as a model of fatigue damage for some ceramic materials [18]. The Walker equation describes the crack growth increment per cycle, n, as

$$\frac{da(\Psi,n)}{dn} = AK_{Iemax}^{N-Q}(\Psi,n)\Delta K_{Ic}^Q(\Psi,n) \quad (15)$$

where

$$K_{Iemax}(\Psi,n) = \sigma_{Ieqcmax}(\Psi,n)Y\sqrt{a(\Psi,n)}$$

and $\Delta K_{Ic}(\Psi,n)$ represents the range of the stress intensity factor over the load cycle. The subscripts max and min indicate the maximum and minimum cycle stress, respectively. The cyclic fatigue parameters A, N, and Q are experimentally determined. The Walker equation reduces to the Paris law [16] when N and Q are equal in value. The integration of Equation (15) parallels that of Equation (2), yielding the cyclic fatigue equivalent stress distribution:

$$\sigma_{Ieqc,0}(\Psi,n_f) = \left[\frac{\int_{n=0}^{n_f} [1 - R(\Psi,n)]^Q \sigma_{Ieqcmax}^N(\Psi,n)dn}{B} + \sigma_{Ieqcmax}^{N-2}(\Psi,n_f) \right]^{1/N-2} \quad (16)$$

where $R(\Psi,n)$ is the ratio of the minimum to maximum cyclic stress, n_f is the number of cycles to failure, and B is now expressed in units of stress^2 × cycle. The parameters B and N are determined from cyclic data.

EVALUATION OF FATIGUE PARAMETERS FROM INHERENTLY FLAWED SPECIMENS

Lifetime reliability of structural ceramic components depends on the history of the loading, the component geometry, the distribution of preexisting flaws, and the parameters that characterize subcritical crack growth. These crack growth parameters must be measured under conditions representative of the service environment. When determining fatigue parameters from rupture data of naturally flawed specimens, the statistical effects of the flaw distribution must be considered, along with the strength degradation effects of subcritical crack growth. In the following discussion, fatigue parameter estimation methods are described for surface flaw analysis using the power law formulation for constant stress rate loading (dynamic fatigue). Analogous formulations for volume flaws, static fatigue, and cyclic fatigue have also been developed [1].

For the uniaxial Weibull distribution, the probability of failure is expressed as

$$P_{fs}(t_f) = 1 - \exp\left[-k_{wS}\int_A \sigma_{1,0}^{m_i}(\Psi)dA\right] \tag{17}$$

where Ψ represents a location (x,y) and $\sigma_{1,0}$ denotes the transformed uniaxial stress analogous to $\sigma_{Ieqc\theta}$ as defined in Equation (13). The Weibull crack density coefficient is given by

$$k_{wS} = \frac{1}{\sigma_{oS}^{m_i}} \tag{18}$$

The Weibull scale parameter, σ_{oS}, corresponds to the stress level where 63.21 percent of specimens with unit area would fail and has units of stress × area$^{1/m_i}$. CARES/LIFE normalizes the various fracture criteria to yield an identical probability of failure for the uniaxial stress state. This is achieved by adjusting the fatigue constant B. For the uniaxial Weibull model, this adjusted value is denoted by B_W, and for the Batdorf model, it is denoted by B_B. From the dynamic fatigue Equation (12), substituting

B_{ws} for B, N_s for N, the uniaxial stress σ_1 for σ_{Ieq}, and rearranging Equation (17) while assuming that

$$\frac{\sigma_1^2(\Psi, t_f) t_f}{(N_s + 1) B_{ws}} \gg 1 \tag{19}$$

the median behavior of the experimental dynamic fatigue data can be described by

$$\sigma_{f0.5} = A_d \dot{\sigma}^{1/(N_s+1)} \tag{20}$$

where $\sigma_{f0.5}$ is the median rupture stress of the specimen and $\dot{\sigma}$ represents the stress rate at the location of maximum stress. The constant A_d is

$$A_d = \left(\frac{(N_s + 1) B_{ws} \sigma_{oS}^{N_s-2}}{\left[\dfrac{A_{ef}}{\left[\ln\left(\dfrac{1}{1 - 0.50} \right) \right]^{1/\tilde{m}_s}} \right]^{1/\tilde{m}_s}} \right)^{1/(N_s+1)} \tag{21}$$

where

$$\tilde{m}_s = \frac{m_s}{N_s - 2}$$

The constants A_d and A_{ef} are obtained by equating risks of rupture. A_{ef} is a modified effective area required for the time-dependent formulation. For the uniaxial Weibull distribution, the expression for the modified effective area is

$$A_{ef} = \int_A \left(\frac{\sigma_1(\Psi, t_f)}{\sigma_f} \right)^{\tilde{m}_s N_s} dA \tag{22}$$

where $\sigma_1(\Psi, t_f)$ denotes the maximum principal stress distribution. For multiaxially stressed components, the Batdorf technique is used to evaluate fatigue parameters. The analogous formulation for A_{ef} is then

$$A_{ef} = \frac{2\bar{k}_{BS}}{\pi} \int_A \left[\int_0^{\pi/2} \left(\frac{\sigma_{Ieq}(\Psi, t_f)}{\sigma_f} \right)^{\tilde{m}_s N_s} d\alpha \right] dA \tag{23}$$

where the normalized Batdorf crack density coefficient is $\bar{k}_{BS} = k_{BS}/k_{wS}$. Equation (21) is applicable, except that B_{BS} replaces B_{wS}. The relationship between B_{BS} and B_{wS} for a uniaxial load is established by equating the risk of rupture of the Batdorf model with that of the uniaxial Weibull model [1]

$$\frac{B_{wS}}{B_{BS}} = \left[\frac{\pi \int_A \sigma^{\bar{m}_sN_S}(\Psi, t_f)dA}{2\bar{k}_{BS} \int_A \int_0^{\pi/2} \sigma_{ieq}^{\bar{m}_sN_S}(\Psi, t_f)d\alpha dA} \right]^{1/\bar{m}_s} \tag{24}$$

As N_S becomes large, Equation (24) approaches unity.

The terms A_d and N_S in Equation (21) are determined from experimental data. Taking the logarithm of Equation (20) yields

$$\ln \sigma_{f0.5} = \ln A_d + \frac{1}{N_S + 1} \ln \dot{\sigma} \tag{25}$$

Linear regression analysis of the experimental data is used to solve Equation (25). The median value method is based on least squares linear regression of median data points for various stress rates. Another technique uses least squares linear regression on all the data points. A third option for estimating fatigue parameters is a modification to a method used by Jakus [25]. In this procedure, fatigue parameters are determined by minimizing the median deviation of the logarithm of the failure stress. The median deviation is the mean of the residuals, where the residual is defined as the absolute value of the difference between the logarithm of the failure stress and the logarithm of the median value. In CARES/LIFE, this minimization is accomplished by maximizing $\bar{m}_s(N_S + 1)$, estimated from the data versus the fatigue exponent.

To obtain A_d based on the median line of the distribution, the following steps are taken. Experimental data at a sufficient number of discrete levels of applied stressing rate are transformed to equivalent failure times t_{Ti} at a fixed stress rate $\dot{\sigma}_T$ [equating failure probabilities using Equation (17)]

$$t_{Ti} = t_{fi}\left(\frac{\dot{\sigma}_i}{\dot{\sigma}_T}\right)^{N_s/(N_s+1)} \tag{26}$$

where the subscript T indicates a transformed value, the subscript i denotes each observed data number, and $t_{fi} = \sigma_{fi}/\dot{\sigma}_i$. In CARES/LIFE the value of $\dot{\sigma}_T$ is taken as the lowest level of stressing rate in the data set.

With the data defined by a single Weibull distribution, parameter estimation is performed on the transformed data using

$$P_{fi} = 1 - \exp\left[-\left(\frac{t_{Ti}}{t_{T\theta s}}\right)^{\bar{m}_s(N_s+1)}\right] \qquad (27)$$

where the characteristic time is

$$t_{T\theta s} = \left[\frac{(N_s + 1)B\sigma_{oS}^{N_s-2}}{A_{ct}^{1/\bar{m}_s}\dot{\sigma}_T^{N_s}}\right]^{1/(N_s+1)} \qquad (28)$$

CARES/LIFE performs least squares or maximum likelihood Weibull parameter estimation as described in Pai [26] to solve Equation (27) for $\bar{m}_s(N_s + 1)$ and $t_{T\theta s}$. Substituting $\dot{\sigma}_T$ for $\dot{\sigma}$ in Equation (20) and solving for the rupture stress in Equation (27), corresponding to a 50% probability of failure, $\sigma_{T0.50}$, yields

$$A_d = t_{T\theta s}\left\{\dot{\sigma}_T^{N_s}\left[\ln\left(\frac{1}{1 - 0.50}\right)\right]^{1/\bar{m}_s}\right\}^{1/(N_s+1)} \qquad (29)$$

This value of A_d is used with the fatigue exponent N_s estimated either with the least squares (using all experimental rupture stresses) or median deviation method. The fatigue constant B is obtained from Equations (21) or (28).

EXAMPLE

This example demonstrates the use of CARES/LIFE to predict the time-dependent reliability of components under multiaxial loads. The data for this example is from experimental work performed by Chao and Shetty [2] on alumina disks and bars exposed to water at room temperature. Chao and Shetty investigated the relationship between stress state and time-dependent strength degradation, specifically to determine if strength degradation due to slow crack growth in biaxial flexure can be predicted from inert fracture stresses and dynamic fatigue assessed in simple uniaxial tests. Time-dependent reliability predictions for the alumina from CARES/LIFE are compared to the results obtained by Chao and Shetty.

Details regarding specimen preparation and testing are given in Reference [2]. Two batches of an alumina ceramic were purchased from the same vendor. The first batch was in the form of plates (127 × 127 × 5 mm) and rods (50.8 mm diameter, 76.2 mm length). The plates and rods were made from the same powder lot with identical isostatic pressing and sintering conditions. The second batch of alumina, in the form of rods, was purchased subsequently to examine dynamic fatigue under biaxial stresses. This material had identical chemistry and preparation as the first batch. The measured properties of the first batch include a fracture toughness, K_k, of 4.13 MPa $\cdot$ $\sqrt{m}$; Young's modulus, E, of 297.2 GPa; and a Poisson's ratio, ν, of 0.23.

Bar specimens were cut from the plate stock, and disks were cut from the rod stock. Specimens were either tested in a dry nitrogen environment at 100 MPa/s to obtain inert (fast-fracture) strengths or in deionized water at various stress rates to obtain the time-dependent fracture strengths. The specimens were carefully prepared to minimize machining damage or failure from edge flaws. The uniaxial specimens were nominally 3 $\times$ 4 $\times$ 45 mm. These specimens were loaded in either four-point flexure, with an outer support span of 40 mm and an inner load span of 20 mm, or in three-point flexure using a 40-mm support span. The disk specimens were nominally 3.175 mm in thickness and 50.8 mm in diameter. They were loaded under uniform pressure on one face and supported on the other face by forty freely rotating ball bearings spaced uniformly along a 49.53-mm diameter circle. Fractography showed that all specimens broke due to a single population of randomly oriented surface flaws.

Table 2.1 lists the fast-fracture Weibull parameters, estimated using the maximum likelihood method, for the various specimen configurations and loads. The values shown correlate very well to those of Reference [2] for the Weibull shape parameter and characteristic strength. The results for the 90% confidence intervals differ somewhat from Reference [2] due to the methods of estimation. CARES/LIFE uses the technique from Thoman, Bain, and Antle [27], while Reference [2] uses a bootstrap technique. The 90% confidence interval on the Weibull modulus significantly overlaps for both bar and disk tests. This is an indication that the strength response is controlled by the same flaw population. Confidence bands on the characteristic strength may only be compared for identical specimen loading and geometry due to the size effect. The confidence intervals obtained for the two disk tests using CARES/LIFE are shown to overlap. The large confidence interval for the batch 2 specimens is due to the small sample size of seven specimens. In Reference [2], the discrepancies in average strength between batch 1 and batch 2 disks were attributed to material processing differences. This will be further discussed.

Fatigue parameters were estimated from four-point bend bar specimens loaded in dynamic fatigue in water. At least fifteen specimens each were tested at stress rates of 0.02, 0.1, 1.0, 10.0, and 100.0 MPa/s. Figure 2.1 shows a plot of the individual failure stresses of the ninety-five specimens tested. Superimposed on this figure are median lines and 5% and 95%

TABLE 2.1. Weibull Parameters Estimated from Inert Data.

Specimen	Weibull Modulus, m_s	90% Confidence Bounds on m_s	Characteristic Strength, $\sigma_{\theta S}$ (MPa)	90% Confidence Bounds on $\sigma_{\theta S}$ (MPa)	Scale Parameter, σ_{oS} (MPa·mm$^{2/m_s}$)
Three-point	25.43	20.47, 29.91	385.9	382.0, 389.9	414.7
Four-point	23.76	19.13, 27.95	353.4	349.6, 357.3	425.8
Disk	22.25	17.12, 26.81	338.7	334.1, 343.5	436.4
Disk (batch 2)	28.98	13.28, 40.88	351.4	341.5, 362.2	423.2

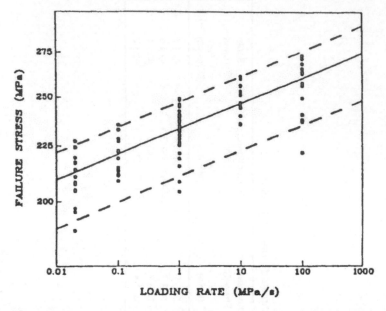

Figure 2.1 Dynamic fatigue of alumina four-point flexure bars in water. Median regression line (solid), and 5% and 95% regression lines (dashed) estimated with tthe median deviation technique.

confidence bounds on this data as estimated with the median deviation technique. Table 2.2 lists the estimated fatigue parameters using the median value, least squares, and median deviation techniques. The median deviation technique shows best agreement with Chao and Shetty for this case. In Reference [2], the power law is used in the following form:

$$\frac{da(t)}{dt} = V_c \left(\frac{K_{\text{Ieq}}}{K_{\text{Ic}}} \right)^N \tag{30}$$

where V_c is the critical crack velocity. The CARES/LIFE program is formulated using the fatigue constant B and, therefore, V_c is not explicitly calculated. However, for comparison, V_c can be computed from CARES/LIFE results using the fracture strength relation [2]

$$\sigma_f = \dot{\sigma} \left[\frac{2K_{\text{Ic}}^{N_S}(N_S + 1)}{V_c Y^{N_S}(N_S - 2)\dot{\sigma}^{N_S}} (a_i^{1-N_S/2} - a_f^{1-N_S/2}) \right]^{1/(N_S+1)} \tag{31}$$

where a_i is initial crack size and a_f is the crack size at failure. The sub-

TABLE 2.2. Fatigue Parameters for Four-Point Bend Bars.

Estimation Method	Fatigue Constant, A_d	Fatigue Exponent, N_S	Fatigue Constant, B_{wS} (MPa²·s)	Fatigue Constant, B_{BS} (MPa²·s)	Crack Velocity, V_c (m·s⁻¹)
Median value	2.339×10^2	36.04	4.590×10^{-1}	4.579×10^{-1}	1.162
Least squares	2.336×10^2	40.84	5.631×10^{-2}	5.617×10^{-2}	8.176
Med deviation	2.337×10^2	41.23	4.783×10^{-2}	4.771×10^{-2}	9.373
Reference [2]	—	40.7	—	—	9.1

script S denotes surface-flaw-dependent properties. Assuming $a_i \gg a_f$ and rearranging the above expression yields

$$\sigma_f = \left[\frac{2K_{Ic}^{N_S}(N_S + 1)a_i^{1-N_S/2}}{V_e Y^{N_S}(N_S - 2)} \right]^{1/(N_f+1)} \dot{\sigma}^{1/(N+1)} \qquad (32)$$

Equating Equations (32) and (20) results in

$$V_e = \frac{2K_{Ic}^{N_S}(N_S + 1)a_i^{1-N_S/2}}{A_d^{N_S+1} Y^{N_S}(N_S - 2)} \qquad (33)$$

The initial crack length a_i and fatigue constant A_d are evaluated here for a failure probability of 50 percent. The crack geometry factor Y for a semi-circular crack is 1.366 [3]. Crack length a_i was determined from Equation (1) using the strength determined from inert testing (Table 2.1) for a 50 percent probability of failure. Chao and Shetty [2] estimated that N_S was 40.7 and V_e was 9.1 m/s for Y equal to 1.24. The differences between the various parameter estimates in CARES/LIFE and Reference [2] are not considered significant. The fatigue velocity V_e is particularly sensitive to N_S as shown in Equation (33). Table 2.2 also lists the fatigue constants B_{wS} and B_{BS}. B_{BS} is determined for a semicircular crack and noncoplanar crack extension with a shear sensitivity constant of $\bar{c} = 0.82$ [3]. The differences between B_{wS} and B_{BS} are small since N_S is relatively large. Finally, the Weibull modulus can be directly estimated from the fatigue data using the relation $\bar{m}_s = m_s/(N_S - 2)$ when N_S and $\bar{m}_s$ are known; for example, using the median deviation method, $m_s = 25.0$, which is consistent with the results shown in Table 2.1.

To obtain disk specimens for dynamic fatigue tests, a second batch of material (of identical dimensions as the first material batch) was secured by Chao and Shetty [2]. Seven of these specimens were broken under inert conditions, and a total of thirty-five specimens were broken at stressing rates of 0.02, 0.1, 10.0, and 100.0 MPa/s. Table 2.1 lists the fast-fracture Weibull parameters of the (batch 2) disks, and Table 2.3 gives the estimates for the fatigue parameters. Table 2.3 shows only B_{BS}, since B_{wS} is formulated only for the uniaxial stress state. Figure 2.2 shows a plot of the individual failure stresses as well as the median line and 5% and 95% confidence bounds estimated with the median deviation technique. One data point was flagged as an outlier (at 100 MPa/s); however, it was not rejected and had little effect on the overall results. Batch 2 material showed a stronger than expected average strength relative to the batch 1 material. Chao and Shetty [2] attributed this to a small increase of K_{Ic} in the material. Although batch 2 was unexpectedly strong, the rate of strength

TABLE 2.3. Fatigue Parameters for Disk (Batch 2) Specimens.

Estimation Method	Fatigue Constant, A_d	Fatigue Exponent, N_s	Fatigue Constant, B_{ss} (MPa²·s)	Crack Velocity, V_c (m·s⁻¹)
Median value	2.293×10^2	35.68	2.675×10^{-1}	1.894
Least squares	2.313×10^2	41.79	3.959×10^{-2}	16.22
Med deviation	2.304×10^2	36.23	2.649×10^{-1}	1.988
Reference [2]	—	36.6	—	2.4

DESIGNING

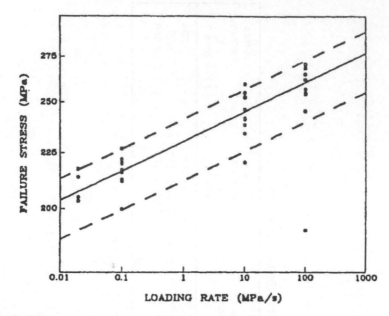

Figure 2.2 Dynamic fatigue of alumina pressure-on-disk flexure specimens in water. Median regression line (solid), 5% and 95% regression lines (dashed) estimated with the median deviation technique.

degradation was similar to that of the batch 1 four-point flexure data. Reference [2] reported that the 90% confidence intervals of N_s for both batches showed significant overlap. This was further confirmed in Table 2.4, which shows the ratio of the natural logarithm of the characteristic strength of the batch 2 disks to the natural logarithm of the characteristic strength of the batch 1 four-point flexure specimens at each stressing rate (maximum likelihood estimates). If the rate of change of strength degrada-

TABLE 2.4. Comparison of Four-Point and Disk (Batch 2) Characteristic Strengths.

Stressing Rate, $\dot{\sigma}$ (MPa·s⁻¹)	Four-Point Strength, $\sigma_{\theta S}$ (MPa)	Disk Strength, $\sigma_{\theta S}$ (MPa)	Strength Ratio, $\dfrac{\ln \sigma_{\theta S_{disk}}}{\ln \sigma_{\theta S_{4-point}}}$
0.020	2.145×10^2	2.116×10^2	0.9975
0.100	2.250×10^2	2.195×10^2	0.9954
10.00	2.531×10^2	2.492×10^2	0.9972
100.0	2.628×10^2	2.620×10^2	0.9995
Inert	3.534×10^2	3.514×10^2	0.9990

tion were stress-state dependent, then the strength ratio would systematically change with stressing rate. However, Table 2.4 shows no such change. Therefore, the differences between fatigue exponents N_s in Table 2.3 and Table 2.2 appear to be reflecting expected statistical variation.

The effect of multiaxial stress states on the material is assessed by comparing the difference in inert strength between the uniaxially loaded four-point bend specimen and the biaxially loaded disk. Assuming that small crack-like imperfections control the failure, the material strength in multiaxial stress states can be correlated to the effects of mixed-mode loading on the individual cracks [8,9]. Shetty [11] developed a simple equation describing the ability of a crack to extend under the combined actions of a normal and shear load on the crack face using an empirically determined parameter, $\bar{c}$. For a semicircular crack, this equation is [3].

$$\sigma_{\text{Ieq}} = \frac{1}{2}\left[\sigma_n + \sqrt{\sigma_n^2 + 3.301\left(\frac{\tau}{\bar{c}}\right)^2}\right] \tag{34}$$

where σ_n and τ are the normal and shear stresses, respectively, acting on the flaw plane.

The failure response of the biaxial flexure specimens and the three-point flexure bars is predicted using the Weibull parameters (Table 2.1) and fatigue parameters (Table 2.2; median deviation results) estimated from the four-point flexure bar rupture tests. The prediction for the three-point bars is compared to experimental results to confirm the expected Weibull size effect. The disk prediction is compared to experimental results in order to make assertions regarding the effect of multiaxial stress states on the fast-fracture and fatigue fracture of the material.

For this analysis, the disk predictions are based on a finite element model of the disk, as shown in Figure 2.3. Brick and wedge solid elements are used to model a 7.5° sector of the disk and appropriate boundary conditions are applied corresponding to a 49.53-mm diameter ring support and a pressure load on one face. A gradient is imposed on the nodal spac-

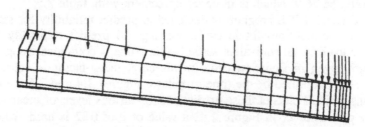

Figure 2.3 Finite element model of a 7.5° sector of the disk. Arrows indicate pressure load on disk.

ing such that the smallest elements have the highest tensile stresses. This is required to obtain a satisfactory convergence of the reliability solution and is independent of the mesh size needed to obtain accurate nodal stresses. To perform surface flaw reliability analysis, quadrilateral and triangular shell elements are attached to the surface nodes on the disk tensile face in order to obtain surface stresses and areas. The shell elements are very thin and have membrane properties only. These elements are such that they contribute negligible stiffness to the model. Verification of the accuracy of the finite-element model is obtained by comparison to available closed form solutions. Chao and Shetty [2] calibrated the applied pressure on the disk to the measured strain at the center of the disk. The strain calibration was used to correlate the fracture stress with the fracture pressure. The CARES/LIFE analysis uses these reported fracture stresses rather than the actual applied pressure on the disk. Reliability predictions for the three-point flexure bar are obtained from closed form solutions of the effective area [1,3].

Probability of failure predictions are made for the disk and three-point bar in the fast-fracture condition and also for dynamic loading at a stressing rate of 1 MPa/s in water. The fracture strength distribution of dynamically (constant stress rate) loaded specimens is characterized by a Weibull distribution with a Weibull modulus, m_{ds}, of

$$m_{ds} = \left(\frac{N_s + 1}{N_s - 2}\right)m_s \tag{35}$$

For this analysis, m_{ds} has a value of 25.58. Table 2.5 lists the estimated Weibull parameters obtained from bars and disks loaded in this condition. In all cases, the estimated Weibull modulus is somewhat higher than predicted; however, the predicted value was within the 90% confidence bounds. Table 2.5 shows that the Weibull modulus did increase from the inert condition, as expected. If the Weibull modulus value of $m_s = 25.0$ obtained directly from the four-point fatigue data were considered, then m_{ds} would be 26.9, which is in better agreement with Table 2.5.

The CARES/LIFE program is designed to predict reliability for static fatigue or constant-amplitude cyclic loading. To predict reliability for dynamic loading, an equivalent static loading time is computed using the approach given in Equation (16). In this case, the g-factor is equal to $1/(N_s + 1)$ multiplied by the time to failure. Figures 2.4 and 2.5 are the resulting Weibull plots of these predictions for various levels of shear sensitivity parameter $\bar{c}$. In Figure 2.4, a value of $\bar{c}$ of 0.82 is used, which corresponds to an approximation of the maximum tangential stress mixed-mode fracture criterion. The dashed line in the figure denotes the results

TABLE 2.5. Weibull Parameters Estimated in Water (1 MPa/s).

Specimen	Weibull Modulus, m_{dS}	90% Confidence Bounds on m_{dS}		Characteristic Strength, σ_{dS} (MPa)	90% Confidence Bounds on σ_{dS} (MPa)	
Three-point	27.83	21.20	33.70	255.6	252.7,	258.5
Four-point	26.70	20.54,	32.17	236.3	233.6,	239.0
Disk	32.66	24.87,	39.53	215.9	213.8,	218.1

from Reference [2]. The small differences indicated are mainly due to the different $\bar{c}$ and crack geometry factor, Y, used and are not due to the different values used for the fatigue parameters. This figure also shows a good correlation between predicted and experimental results for the three-point flexure bars, which confirms the size effect expected in fast fracture and fatigue. The CARES/LIFE prediction lines for the three-point flexure bars superimpose on the three-point bar prediction lines of Reference [2].

Figure 2.5 shows predictions for $\bar{c}$ values of 0.82, 0.90, and $\gg 1$ (solid, long dashed, and short dashed line, respectively). Since, in this example, the CARES/LIFE fracture predictions are normalized to the (uniaxial stress state) four-point flexure data, then the choice of a fracture criterion only affects the predictions for the (biaxial stress state) disks. The $\bar{c}$ values of 0.82 and much greater than one represent the extreme bounds on the expected mixed-mode shear sensitivity of the flaws in the material. The value of $\bar{c} \gg 1.0$ represents a shear-insensitive fracture criterion, while $\bar{c} = 0.82$ is highly shear sensitive. The 0.90 value represents a choice of $\bar{c}$ that best fits to both inert and fatigue disk data. Optimizing $\bar{c}$ in this manner was not considered in Reference [2]. From this figure, stress state

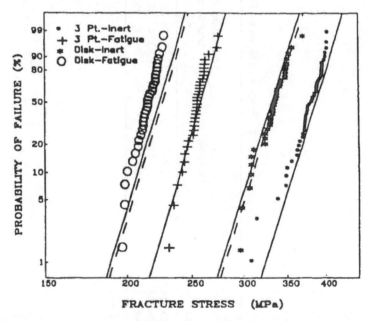

Figure 2.4 Weibull plot showing fast-fracture three-point and disk specimen strengths measured in an inert environment, as well as three-point and disk specimen fracture strengths dynamically loaded at a rate of 1 MPa/s. Solid lines are corresponding CARES/LIFE predictions for $\bar{c} = 0.82$. Dashed lines are corresponding predictions from Reference [2].

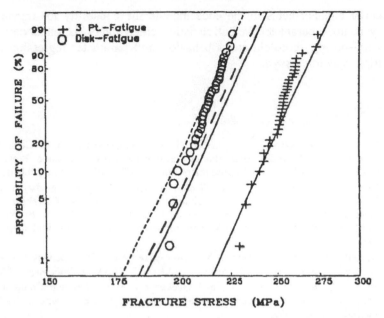

Figure 2.5 Weibull plot showing three-point and disk specimen dynamic fatigue strengths that were loaded at a rate of 1 MPa/s. Solid line corresponds to CARES/LIFE prediction for $\bar{c} = 0.82$ (shear-sensitive criterion). Long dashed line corresponds to the disk prediction for $\bar{c} = 0.90$ and the short dashed line corresponds to the disk prediction for $\bar{c} > 1.0$ (shear-insensitive criterion).

effects are adequately accounted for, both in fast-fracture and subcritical crack growth conditions.

CONCLUSION

The use of structural ceramics for high-temperature applications depends on the strength, toughness, and reliability of these materials. Ceramic components can be designed for service if the factors that cause material failure are accounted for. This design methodology must combine the statistical nature of strength-controlling flaws with fracture mechanics to allow for multiaxial stress states, concurrent flaw populations, and subcritical crack growth. This has been accomplished with the CARES/LIFE public domain computer program for predicting the time-dependent reliability of monolithic structural ceramic components. An example has been given to illustrate the use of this design methodology in CARES/LIFE for predicting the effects of component size, stress distribution, stress state, and subcritical crack growth on the lifetime reliability.

Potential enhancements to the code include the capability for transient analysis, three-parameter Weibull statistics, creep and oxidation modeling, flaw anisotropy, threshold stress behavior, and parameter regression for multiple specimen sizes.

REFERENCES

1 Nemeth, N. N., Powers, L. M., Janosik, L. A., and Gyekenyesi, J. P. "Ceramics Analysis and Reliability Evaluation of Structures LIFE Prediction Program (CARES/LIFE) Users and Programmers Manual," TM-106316, to be published.

2 Chao, L. Y. and Shetty, D. K. "Time-Dependent Strength Degradation and Reliability of an Alumina Ceramic Subjected to Biaxial Flexure," *Life Prediction Methodologies and Data for Ceramic Materials, ASTM STP 1201*, C. R. Brinkman and S. F. Duffy, eds., American Society for Testing and Materials, Philadelphia, 1993.

3 Nemeth, N. N., Manderscheid, J. M., and Gyekenyesi, J. P. "Ceramics Analysis and Reliability Evaluation of Structures (CARES)," NASA TP-2916, Aug. 1990.

4 Powers, L. M., Starlinger, A., and Gyekenyesi, J. P. "Ceramic Component Reliability with the Restructured NASA/CARES Computer Program," NASA TM-105856, Sept. 1992.

5 Barnett, R. L., Connors, C. L., Hermann, P. C., and Wingfield, J. R. "Fracture of Brittle Materials under Transient Mechanical and Thermal Loading," U.S. Air Force Flight Dynamics Laboratory, AFFDL-TR-66-220, 1967 (NTIS AD-649978).

6 Freudenthal, A. M. "Statistical Approach to Brittle Fracture," *Fracture, Vol. 2: An Advanced Treatise, Mathematical Fundamentals*, H. Liebowitz, ed., Academic Press, 1968, pp. 591-619.

7 Weibull, W. A. "The Phenomenon of Rupture in Solids," *Ingenoirs Vetenskaps Akadanien Handlinger*, 1939, No. 153.

8 Batdorf, S. B. and Crose, J. G. "A Statistical Theory for the Fracture of Brittle Structures Subjected to Nonuniform Polyaxial Stresses," *Journal of Applied Mechanics*, Vol. 41, No. 2, 1974, pp. 459-464.

9 Batdorf, S. B. and Heinisch, H. L., Jr. "Weakest Link Theory Reformulated for Arbitrary Fracture Criterion," *Journal of the American Ceramic Society*, Vol. 61, No. 7-8, 1978, pp. 355-358.

10 Palaniswamy, K. and Knauss, W. G. "On the Problem of Crack Extension in Brittle Solids under General Loading," *Mechanics Today*, Vol. 4, 1978, pp. 87-148.

11 Shetty, D. K. "Mixed-Mode Fracture Criteria for Reliability Analysis and Design with Structural Ceramics," *Journal of Engineering for Gas Turbines and Power*, Vol. 109, No. 3, 1987, pp. 282-289.

12 Evans, A. G. "A General Approach for the Statistical Analysis of Multiaxial Fracture," *Journal of the American Ceramic Society*, Vol. 61, 1978, pp. 302-306.

13 Matsuo, Y. *Trans. of the Japan Society of Mechanical Engineers*, Vol. 46, 1980, pp. 605-611.

14 Evans, A. G. and Wiederhorn, S. M. "Crack Propagation and Failure Prediction in Silicon Nitride at Elevated Temperatures," *Journal of Material Science*, Vol. 9, 1974, pp. 270-278.

15 Wiederhorn, S. M. *Fracture Mechanics of Ceramics*. R. C. Bradt, D. P. Hasselman, and F. F. Lange, eds., Plenum, New York, 1974, pp. 613–646.

16 Paris, P. and Erdogan, F. "A Critical Analysis of Crack Propagation Laws," *Journal of Basic Engineering*, Vol. 85, 1963, pp. 528–534.

17 Walker, K. "The Effect of Stress Ratio during Crack Propagation and Fatigue for 2024-T3 and 7075-T6 Aluminum," *ASTM STP 462*, 1970, pp. 1–14.

18 Dauskardt, R. H., James, M. R., Porter, J. R., and Ritchie, R. O. "Cyclic Fatigue Crack Growth in SiC-Whisker-Reinforced Alumina Ceramic Composite: Long and Small Crack Behavior," *Journal of the American Ceramic Society*, Vol. 75, No. 4, 1992, pp. 759–771.

19 Paris, P. C. and Sih, G. C. "Stress Analysis of Cracks," *ASTM STP 381*, 1965, pp. 30–83.

20 Thiemeier, T. "Lebensdauervorhersage für keramische Bauteile unter mehrachsiger Beanspruchung," Ph.D. dissertation, University of Karlesruhe, Germany, 1989.

21 Sturmer, G., Schulz, A., and Wittig, S. "Lifetime Prediction for Ceramic Gas Turbine Components," ASME Preprint 91-GT-96, June 3–6, 1991.

22 Batdorf, S. B. "Fundamentals of the Statistical Theory of Fracture," *Fracture Mechanics of Ceramics, Vol. 3*, R. C. Bradt, D. P. H. Hasselman, and F. F. Lange, eds., Plenum Press, New York, 1978, pp. 1–30.

23 Mencik, J. "Rationalized Load and Lifetime of Brittle Materials," *Communications of the American Ceramic Society*, March 1984, pp. C37–C40.

24 Evans, A. G. "Fatigue in Ceramics," *International Journal of Fracture*, Dec. 1980, pp. 485–498.

25 Jakus, K., Coyne, D. C., and Ritter, J. E. "Analysis of Fatigue Data for Lifetime Predictions for Ceramic Materials," *Journal of Material Science*, Vol. 13, 1978, pp. 2071–2080.

26 Pai, S. S. and Gyekenyesi, J. P. "Calculation of the Weibull Strength Parameters and Batdorf Flaw Density Constants for Volume and Surface-Flaw-Induced Fracture in Ceramics," NASA TM-100890, 1988.

27 Thoman, D. R., Bain, L. J., and Antle, C. E. "Inferences on the Parameters of the Weibull Distribution," *Technometrics*, Vol. 11, No. 3, 1969, pp. 445–460.

CHAPTER 3

Probabilistic Micromechanics and Macromechanics of Polymer Matrix Composites

G. T. MASE,[1]
P. L. N. MURTHY[2] and C. C. CHAMIS[2]

OVERVIEW

A probabilistic evaluation of an eight-ply graphite/epoxy quasi-isotropic laminate was completed using the integrated composite analyzer (ICAN) in conjunction with Monte Carlo simulation and fast probability integration (FPI) techniques. Probabilistic input included fiber and matrix properties, fiber misalignment, fiber volume ratio, void volume ratio, ply thickness, and ply layup angle. Cumulative distribution functions (CDFs) for select laminate properties are given. To reduce a number of simulations, a fast probability integration (FPI) technique was used to generate CDFs for the select properties in the absence of fiber misalignment. These CDFs were compared to a second Monte Carlo simulation done without fiber misalignment effects. It is found that FPI requires a substantially less number of simulations to obtain the cumulative distribution functions as opposed to Monte Carlo Simulation techniques. Furthermore, FPI provides valuable information regarding the sensitivities of composite properties to the constituent properties, fiber volume ratio, and void volume ratio.

NOMENCLATURE

E_{cxx} composite elastic modulus (MPSI) about structural axes
$E_{f11}, E_{f22}, G_{f12}$ fiber elastic moduli about material axes

[1]GMI Engineering and Management Institute, Department of Mechanical Engineering, Flint, MI, U.S.A.
[2]NASA Lewis Research Center, Cleveland, OH, U.S.A.

E_m, G_m matrix elastic moduli

$F(x), p_f$ cumulative distribution function

$f(x)$ probability density function

$f_{\underline{z}}(\underline{X})$ joint probability density function

G_{cxy} composite shear modulus (Mpsi) about structural axes

g FPI limit state function

g_1 linear approximation of g

g_2 incomplete quadratic approximation of g

k_f, k_m, k_v fiber volume fraction, matrix volume fraction, void volume fraction

u standardized normal deviate

x uniform deviate

y normal, Weibull, or gamma deviate

Z FPI response function

GREEK LETTERS

α, β Weibull distribution parameters

α_{cxx} composite thermal expansion coefficient (ppm/°F) about structural axes

$\Gamma(x)$ gamma function

λ, x gamma distribution parameters

μ, σ normal distribution parameters

ν_m fiber and matrix Poisson's ratios

INTRODUCTION

The properties of the polymer matrix composites display considerable scatter because of the variation inherent in the properties of constituent materials. Distinct distributions to describe the effects of scatter on composite properties facilitate the composite mechanics calculations. For example, composite strength is often examined probabilistically by assuming that the ply failure strength has a specific distribution (usually Weibull), which is then used in a laminate failure criterion [1]. Analysis of this type has the shortcoming that different failure mechanisms occurring at a lower level, that is, at the fiber and matrix level are not directly accounted for when the ply failure stress is the primitive random variable.

A better approach to quantify the uncertainties in the behavior of composites would be to account for the variations in the properties starting from the constituent (fiber and matrix) level and integrating progressively to arrive at the global or composite level behavior. Typically, these uncertainties may occur at the constituent level (fiber and matrix properties), at the ply level (fiber volume ratio, void volume ratio, etc.), and the composite level (ply angle and layup). In this chapter, a computational simulation technique is described, which accounts for uncertainties at various levels to predict the behavior of a quasi-isotropic graphite/epoxy (0/45/90/ −45), laminate.

MICROMECHANICAL AND MACROMECHANICAL UNCERTAINTIES

UNCERTAINTIES AT THE MICROMECHANICS LEVEL

To account for uncertainties at all levels of a composite, one has to start with uncertainties at the fiber and matrix level and use composite mechanics to obtain laminate level response. In the present effort, the composite mechanics available in ICAN [2] are utilized to obtain the ply/laminate level response. At the micromechanics level, twenty-nine parameters (the constituent properties) are required by ICAN as input [2] (Figure 3.1). In addition, three fabrication process variables (Table 3.2) are needed to compute ply properties. For the most part, these properties were considered to be normally distributed about some mean value. However, the fiber and matrix strengths were taken to be distributed as a Weibull distribution, which is widely accepted for strength distributions because of its dispersed left tail and sharp right tail, which represent experimental data well.

The distribution types and parameters for the fiber and matrix constituent material properties are given in Table 3.1. Using the Monte Carlo simulation, these distribution types reproduce histogram (frequency of occurrence) plots, as shown in Figure 3.2 for fiber longitudinal modulus and Figure 3.3 for fiber longitudinal strength. It would require testing of 1000 specimens to generate them experimentally (a rather expensive and time-consuming task).

It is worth noting that, for the Weibull distribution, the mean is not parameter 1 nor is the variance parameter 2. In this case, the probability density function, mean, and variances are given by Reference [3].

$$f(y) = \frac{\beta}{\lambda} \left(\frac{y}{\lambda}\right)^{\beta-1} e^{-(y/\lambda)^{\beta}}$$

where

$$\lambda^{\beta} = \frac{1}{\alpha} \tag{1}$$

$$\text{Mean} = \alpha^{-1/\beta}\Gamma\left(1 + \frac{1}{\beta}\right), \quad \text{Variance} = \alpha^{-2/\beta}\left[\Gamma\left(1 + \frac{2}{\beta}\right) - \Gamma^2\left(1 + \frac{1}{\beta}\right)\right]$$

$$\tag{2}$$

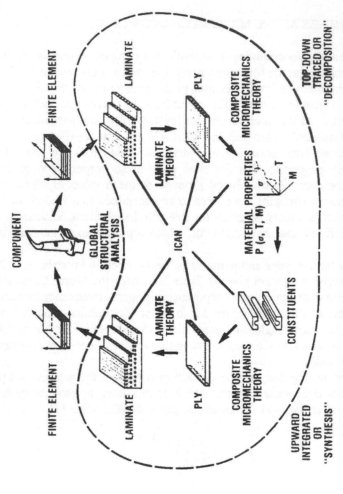

Figure 3.1 Integrated composite micro- and macromechanics analysis embedded in the computer code ICAN.

TABLE 3.1. Constituent Input Distribution Parameters for ICAN.

| | Distribution | | Parameter 1 | Parameter 2 |
	Units	Type		
E_{f11}	Mpsi	Normal	$\mu = 31.0$	$\sigma = 1.5$
E_{f22}	Mpsi		$\mu = 2.0$	$\sigma = 0.10$
G_{f12}	Mpsi		$\mu = 2.0$	$\sigma = 0.10$
G_{f23}	Mpsi		$\mu = 1.0$	$\sigma = 0.05$
ν_{f12}	in./in.		$\mu = 0.20$	$\sigma = 0.01$
ν_{f23}	in./in.		$\mu = 0.25$	
α_{f11}	ppm/°F		$\mu = 0.2$	
α_{f22}	ppm/°F		$\mu = 0.2$	▼
ϱ_f	lb/in.³	▼	$\mu = 0.063$	$\sigma = 0.003$
N_f	—	Fixed	$\mu = 10{,}000$	$\sigma = 0$
d_f	in.	Normal	$\mu = 0.003$	$\sigma = 0.00015$
C_f	Btu/lb		$\mu = 0.20$	$\sigma = 0.01$
K_{f11}	a		$\mu = 580$	$\sigma = 2.9$
K_{f22}	a		$\mu = 58$	$\sigma = 2.9$
K_{f33}	a	▼	$\mu = 58$	$\sigma = 2.9$
S_{fT}	ksi	Weibull	$\beta = 400$	$\sigma = 40$
S_{fC}	ksi	Weibull	$\beta = 400$	$\sigma = 40$
E_m	Mpsi	Normal	$\mu = 0.500$	$\sigma = 0.025$
G_m	Mpsi	b	—	—
ν_m	in./in.	Normal	$\mu = 0.35$	$\sigma = 0.035$
α_m	ppm/°F		$\mu = 36$	$\sigma = 4$
ϱ_m	lb/in.³		$\mu = 0.0443$	$\sigma = 0.0022$
C_m	Btu/lb		$\mu = 0.25$	$\sigma = 0.0125$
K_m	a	▼	$\mu = 1.25$	$\sigma = 0.06$
S_{mT}	ksi	Weibull	$\mu = 15$	$\alpha = 5$
S_{mC}	ksi	Weibull	$\mu = 35$	$\alpha = 20$
S_{mS}	ksi	Weibull	$\mu = 13$	$\alpha = 7$
β_m	in./in. 1% moisture	Normal	$\mu = 0.004$	$\sigma = 0.0002$
D_m	in.²/sec	Normal	$\mu = 0.002$	$\sigma = 0.0001$

[a]Btu·in./hr/ft²/°F
[b]G_m is calculated using E_m and ν_m and isotropy.

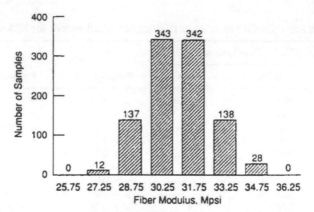

Figure 3.2 Monte Carlo simulation (1000 samples) of fiber longitudinal modulus from a normal distribution.

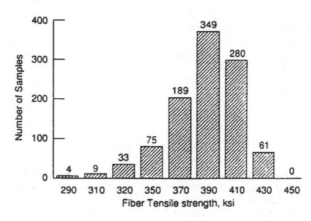

Figure 3.3 Monte Carlo simulation (1000 samples) of fiber longitudinal strength from a Weibull distribution.

UNCERTAINTIES AT THE PLY LEVEL

The next level of uncertainties enters at the ply level. A typical graphite fiber has a nominal diameter of 0.0003 in., which means that a single ply contains many fibers through the thickness. If an eight-ply graphite/epoxy composite is considered with a nominal thickness of 0.04 in., each ply will be approximately 0.005 in. thick. Taking into account an interfiber spacing of 0.00005 in. (for a ply with fiber volume ratio 0.6), there are about fifteen fibers through the thickness of each ply. All of these fibers will have a certain amount of misalignment (random orientation). To account for this randomness in probabilistic micromechanics, linear laminate theory is used where each ply is broken down (substructured into fifteen subplies [2]). Each of these subplies was assumed to be normally distributed about the fiber direction with fiber orientations lying within ±5° of the 0°-ply direction. The properties of the constituents are assumed to be the same in the subplies within each ply. The fiber volume and ply thickness were represented as normally distributed while the void volume was represented as a gamma distribution. A gamma distribution was the proper choice for the void volume ratio because there is no probability for zero void volume and a bias towards higher void volumes.

As was the case for the Weibull distribution, the parameters given for the gamma distribution do not directly represent the mean and variance of the distribution. The probability density function, mean, and variance for the gamma function are given by [3]:

$$f(y) = \frac{\lambda^x}{\Gamma(k)} e^{-\lambda y} y^{k-1} \tag{3}$$

$$\text{Mean} = \frac{k}{\lambda}, \quad \text{Variance} = \frac{k}{\lambda^2} \tag{4}$$

The ply level distribution parameters are given in Table 3.2.

TABLE 3.2. Ply Input Distribution Parameters for ICAN.

	Units	Distribution Type	Parameter 1	Parameter 2
k_f	Percent	Normal	$\mu = 60$	$\sigma = 3$
k_v	Percent	Gamma	$\lambda = 2$	$k = 6$
Θ_f	Degrees	Normal	$\mu = 0$	$\sigma = 3.33$

TABLE 3.3. Laminate Input Distribution Parameters for ICAN.

	Units	Distribution Type	Parameter 1	Parameter 2
Θ_l	Degrees	Normal	$\mu = 0$	$\sigma = 3.33$
t_l	Inches	Normal	$\mu = t_0$	$\sigma = 0.05 t_0$

UNCERTAINTIES AT THE LAMINATE LEVEL

The uncertainty considered at the laminate level was that of ply orientation and thickness. Each ply in the $(0/45/90/-45)_s$ was given a normal distribution with a $3.33°$ standard deviation about the deterministic angle (Table 3.3).

MONTE CARLO SIMULATION

Given the distribution of Tables 3.1, 3.2, and 3.3 for fiber and matrix properties, ply and laminate inputs, the uncertainties in the composite properties need to be quantified with appropriate cumulative distribution functions (CDFs). One approach to achieve this is to use a Monte Carlo simulation technique. The first step in this process involves running ICAN with randomly selected input variables from the predetermined probability distribution functions many times. The output comprised of the composite properties is saved. The second step consists of processing the various property output data to compute the desired CDF. An obvious disadvantage of such an approach is the enormous number of output sets that must be obtained to get reasonable accuracy in the output CDFs.

GENERATION OF NORMAL DISTRIBUTIONS

In order to generate the input distributions, a uniform deviate (random number between 0 and 1) must first be generated. Rather than use a machine routine to generate the random number, a portable (machine independent) uniform deviate routine from Press et al. [4] was used, which was based on a three linear congruential generator method. This routine also had the advantage of being able to reinitialize the random sequence.

The uniform deviate was used to generate a normal deviate by the Box-Muller method [4]. With the aid of the normal distribution $f(y)$ given by

$$f(y)dy = \frac{1}{\sqrt{2\pi}} e^{-y^2/2} dy \tag{5}$$

and the transformation between uniform deviates x_1, x_2 and the variables y_1, y_2 are given by

$$y_1 = \sqrt{-2 \ln x_1} \cos 2\pi x_2, \quad y_2 = \sqrt{-2 \ln x_1} \sin 2\pi x_2 \qquad (6)$$

the inverse transformation can be written as

$$x_1 = e^{(y_1^2 + y_2^2)/2}, \quad s_2 = \frac{1}{2\pi} \arctan \frac{y_2}{y_1} \qquad (7)$$

The Jacobian of this transformation is

$$\frac{\partial(x_1, x_2)}{\partial(y_1, y_2)} = -\left[\frac{1}{\sqrt{2\pi}} e^{-(y_1^2)/2} \right]\left[\frac{1}{\sqrt{2\pi}} e^{-(y_2^2)/2} \right] \qquad (8)$$

which shows that each y is distributed normally. This shows that Equation (6) leads to an explicit formula for calculating a normal deviate.

GENERATION OF WEIBULL DISTRIBUTION

To generate a Weibull distribution from a uniform deviate, one can integrate the probability density function and then solve for the Weibull deviate. This gives

$$y = \beta[-\ln(1-x)]^{1/\alpha} \qquad (9)$$

as a point from the Weibull distribution where x is a uniform deviate.

GENERATION OF GAMMA DISTRIBUTION

To generate a deviate from a gamma distribution, a uniform deviate, x, was taken and then the zero of the function

$$w(y) = \sum_0^\infty \frac{\lambda^k}{\Gamma(k)} e^{-\lambda y} y^{k-1} dx - x \qquad (10)$$

was found. This was numerically inefficient because it involved numerical integration and root finding by the bisection method, but the program ran with sufficient speed to overlook this fact.

The program ICAN was modified so that the properties shown in Tables 3.1, 3.2, and 3.3 were given a value from their respective distributions.

Output for the layer and composite properties were saved for 200 samples. While 200 samples is probably not enough to converge to the actual CDF, the results do show a good qualitative trend. The cumulative distribution functions were constructed from these samples for three typical composite properties. The selected composite properties are the composite longitudinal modulus E_{cxx}, the composite compressive strength S_{cxx}, and the composite thermal expansion coefficient α_{cxx}. These CDFs are shown in Figures 3.4 to 3.6. By its symmetry, the CDF of E_{cxx} appears to be normally distributed, while S_{cxx} exhibits a Weibull shape [5].

FPI SIMULATION

An alternative approach to obtain the required cumulative distribution functions is to use fast probability integration (FPI) program [6]. FPI helps generate the required CDFs quicker, with reasonable accuracy and a fewer number of sample output data. Also, it generates more information than what can be expected from a Monte Carlo simulation. The additional information that FPI offers is the output variable sensitivity information based on the probabilistic inputs.

A brief overview of FPI is given below. The reader is advised to refer to Reference [6] for a detailed discussion.

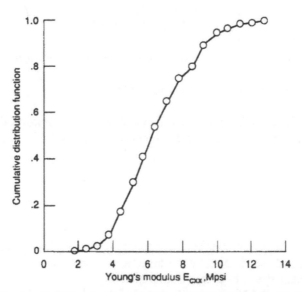

Figure 3.4 Cumulative distribution function for composite ([0/45/90/−45]$_s$) modulus E_{cxx} from Monte Carlo simulation (200 samples), which includes ply substructuring effects.

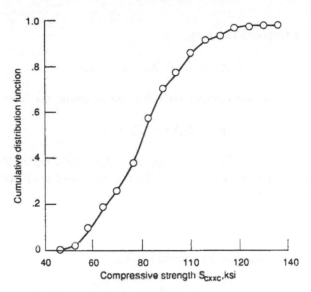

Figure 3.5 Cumulative distribution function for composite ([0/45/90/−45]ₛ) modulus S_{cxx} from Monte Carlo simulation (200 samples), which includes ply substructuring effects.

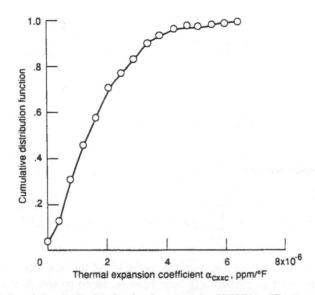

Figure 3.6 Cumulative distribution function for composite ([0/45/90/−45]ₛ) thermal expansion coefficient α from Monte Carlo simulation (200 samples), which includes ply substructuring effects.

Consider a response function

$$Z(\underline{X}) = Z(X_1, X_2, \ldots, X_n) \tag{11}$$

where $X_1, \ldots, X_n$ are random variables. Also, define the function

$$g = Z(\underline{X}) - Z_0 = 0 \tag{12}$$

as the limit state, with Z_0 a real value of $Z(\underline{X})$. The CDF of Z at Z_0 is equal to the probability that $g \leq 0$. If the probability of a desired output, p_f, is defined by

$$p_f = P[g < 0] \tag{13}$$

an exact solution of p_f can be obtained from

$$p_f = \ldots \int_\Omega f_{\underline{x}}(\underline{X}) d\underline{X} \tag{14}$$

where $f_{\underline{x}}(\underline{X})$ is the joint probability density function and Ω is the region defined by $g \leq 0$.

The evaluation of the preceding integral is often intractable, and this leads to the need for an approximate method of evaluation p_f. In doing this, FPI approximates the function g using a Taylor's series expansion as a linear

$$g_1(\underline{u}) = a_0 + \sum_{i=1}^{n} a_i(u_i - u_i^*) \tag{15}$$

or incomplete quadratic

$$g_2(\underline{u}) = a_0 + \sum_{i=1}^{n} a_i(u_i - u_i^*) = \sum_{i=1}^{n} b_i(u_i - u_i^*)^2 \tag{16}$$

function where u_i^* is the most probable point [6] of the random variable u_i. Note that the random variables $\underline{X}$ have been replaced by standardized normal variables $\underline{u}$. The coefficients of these expansions are obtained numerically, and then the probability $P[g < 0]$ is computed.

Because of the approximate form of the g-function, FPI requires at least $n + 1$ or $2n + 1$ data sets to evaluate the linear or quadratic g-function coefficients a_0, a_i, and b_i from which the probability is found. In the present effort, only the ply level variations in the properties (twenty-nine), fiber volume ratio, and void volume ratio are considered as random variables. This means that at least thirty-two (twenty-nine constituent properties, fiber volume ratio, void volume ratio $+ 1$) ICAN runs are needed for the linear approximation and sixty-three for the quadratic approximation. A typical data set to FPI consisted of ICAN run with one perturbed independent variable while all others remain at mean value. For the linear case, the variable was perturbed one standard deviation from its mean value. In the quadratic case, the independent variables were perturbed twice, one standard deviation each, on both sides of the mean value.

Three typical composite properties E_{cxx}, G_{cxy}, and α_{cxx} were chosen as the output variables for the study. Since the goal was to have a minimum number of ICAN runs, the CDF of E_{cxx} was calculated using thirty-two data sets with linear FPI analysis and 125 data sets with quadratic analysis. With these two cases, the CDFs computed by FPI lay on top of each other, indicating that thirty-two data sets will give a good approximation for the CDF of E_{cxx}.

To identify the computational savings that FPI has over a Monte Carlo simulation, the CDFs for E_{cxx}, G_{cxy}, and α_{cxx} were compared for a thirty-two–sample FPI case with a thirty-one– and 90-sample Monte Carlo simulation (Figures 3.7 to 3.9). It appears that the Monte Carlo simulation is converging to the FPI simulation, but the FPI simulation only needed thirty-two samples.

FPI SENSITIVITY OUTPUT

As was previously mentioned, one advantage of FPI is the sensitivity data that is produced. Before the actual sensitivity numbers are given for the composite modulus E_{cxx}, it will be helpful to examine how this quantity is calculated by ICAN.

The modulus E_{cxx} is the $(1, 1)$ entry of the matrix (2)

$$[E_c] = \frac{1}{t_c}\left[\sum_{i=1}^{N_l} (Z_{1i} - Z_{1i})([R_1]^T[E_1][R_1])_i \right] \tag{17}$$

where t_c is the thickness of the composite; Z_{1i} is the distance from the bottom of the composite to the ply; $[R_1]$ is rotation matrix, which is a function of the ply angle; $[E_1]$ is the matrix of the layer elastic constants, distortional energy coefficient. To calculate the ply elastic moduli matrix, $[E_1]$, the components are calculated from primitive variables E_{f11}, E_{f22}, E_m, G_{f12}, G_{f23}, G_m, k_f, k_m, ν_{f12}, and ν_m [2]. So in the case of no ply substructur-

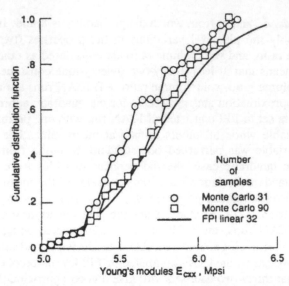

Figure 3.7 Cumulative distribution function for composite ([0/45/90/ −45]ₛ) modulus E_{cxx} simulation with Monte Carlo (no ply substructuring) and fast probability integration (FPI).

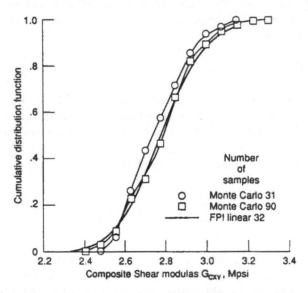

Figure 3.8 Cumulative distribution function for composite ([0/45/90/ −45]ₛ) Shear modulus G_{cxy} simulation with Monte Carlo (no ply substructuring) and fast probability integration (FPI).

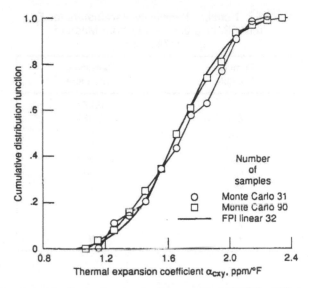

Figure 3.9 Cumulative distribution function for the composite ([0/45/90/−45]$_s$) thermal expansion α_{cxx} simulated with Monte Carlo (no ply substructuring) and fast probability integration (FPI).

ing or ply angle variation, the composite modulus should be a function of only those ten primitive variables. It is noted that k_m is calculated by

$$k_m = 1 - k_f - k_v \qquad (18)$$

and is not listed as a primitive variable. Thus, k_v will be used instead of k_m in the sensitivity analysis.

For the input random variables given previously, FPI calculated a mean E_{cxx} of $\mu = 5.744$ Mpsi with a standard deviation of $\sigma = 0.363$ Mpsi. The sensitivities at $\pm 0.3\sigma$ are given in Table 3.4. As would be expected, the most sensitive primitive variables are the fiber modulus E_{f11} and the fiber volume ratio k_f. Primitive variable G_{f23} has a zero sensitivity, which is consistent with the definition of E_{cxx} given in matrix Equation (17).

CONCLUSIONS

A probabilistic evaluation of an eight-ply quasi-isotropic graphite/epoxy [0/45/90/−45]$_s$ laminate was completed using two approaches. The first approach was to use a Monte Carlo simulation technique. The second approach was to use fast probability integration technique (FPI). Probabilistic inputs for this study included constituent micromechanical proper-

TABLE 3.4. Nonzero Sensitivity Parameters for E_{cxx} from FPI at $\pm 0.3\ \sigma$ away from Mean of $\mu = 5.744$ Mpsi.

Primitive Variable	Sensitivity Parameter
k_f	0.778
E_{f11}	.624
E_{f22}	.260
G_{f12}	.130
G_m	.060
E_m	.036
G_{f23}	.0
ν_{f23}	
ν_{f12}	
ν_m	
k_v	▼

ties, fiber misalignment within a ply, fiber volume fraction, void volume percent, and ply angle misalignment for the laminate.

It was demonstrated that the use of the FPI program can greatly reduce the computations needed to generate composite CDFs. FPI was demonstrated by generating CDFs for E_{cxx}, G_{cxy}, and α_{cxx} for a graphite/epoxy $[0/45/90/-45]_s$ composite in the absence of fiber misalignment.

The results of this investigation indicate that an integrated program combining ICAN and FPI is feasible. Such an integrated program offers the potential for a computational efficient probabilistic composite mechanics methodology.

REFERENCES

1 Duva, J. M., Lang, E. J., Mirzadeh, F., and Herakovich, C. T. "A Probabilistic Perspective on the Failure of Composite Laminate," presented at the *XI U.S. National Congress of Applied Mechanics*, Tucson, AZ, May 1990.

2 Murthy, P. L. N. and Chamis, C. C. "Integrated Composite Analyzer (ICAN), Users and Programmers Manual," NASA TP-2515, 1986.

3 Mood, A. M., Graybill, F. A., and Boes, D. C. *Introduction to the Theory of Statistics.* 3rd Edition, McGraw-Hill, New York, NY, 1986.

4 Press, W. H., Flannery, B. P., Teukolsky, S. A., and Vetterling, W. T. *Numerical Recipes.* Cambridge University Press, Cambridge, U.K., 1986.

5 Stock, T. A. "Probabilistic Fiber Composite Micromechanics," M.S. Thesis, Cleveland State University, Cleveland, OH, 1987.

6 "Probabilistic Structural Analysis Methods (PSAM) for Select Space Propulsion Systems Components, NESSUS/FPI Theoretical Manual," Southwest Research Institute, NASA Contract NAS3-24389, Dec. 1989.

PART 2

THERMOMECHANICAL TESTING

Burner Rig Thermal Fatigue of SiC Continuous Fiber Silicon Nitride Composites under Constant Applied Stress

TURGAY ERTÜRK[1]

OVERVIEW

THE burner rig thermal fatigue behavior of 0/90 continuous SiC fiber Si_3N_4 ceramic composites was investigated under impinged jet fuel flame, a constant applied tensile stress and thermal cycling in the temperature range 500–1350°C. Nine SCS-9 and four SCS-6 SiC fiber-reinforced specimens were thermal fatigued at the NASA Lewis Mach 0.3 atmospheric pressure burner rig test station. The hot pressed $[(0/90)_3/\bar{0}]_s$ SCS-6® or $[(0/90)_s/\bar{0}]_s$ SCS-9® SiC fiber crossply specimens had a fiber volume fraction of 0.30 and a thickness of 3.175 mm. The thermal cycles consisted of one-minute flame impingement (forced convection and radiation heating) followed by one-minute ambient temperature (radiation and natural convection) cooling. Specimens having SCS-6 SiC continuous fibers failed outside the flame impinged zone by high thermal stresses due to steep temperature gradients in twenty to forty cycles at constant applied stresses as low as 110 MPa. SCS-9 specimens, on the other hand, failed at the center of the flame-impinged zone where high-temperature gradients did not exist. The SCS-9 SiC fiber-reinforced material showed runout conditions (1000 cycles without failure) up to 125 MPa constant applied stress. The number of cycles-to-failure declined abruptly at 140 MPa constant applied stress; at 168 MPa, it was 5–42. Surface analysis of the specimens using SIMS and XPS showed no significant chemical modification. No strength or modulus degradation was observed in the SCS-9 runout specimens.

[1]University of Massachusetts Lowell, Department of Chemical and Nuclear Engineering, Lowell, MA, U.S.A.

INTRODUCTION

Crystalline ceramics are ideal materials for high-temperature applications because of their high melting point and slow diffusion kinetics. With advances in the understanding of their mechanical properties [1] and the role of interfaces on composite toughness [2], continuous fiber ceramic composites (CFCCs) have the potential of replacing nickel-based superalloys and other high-temperature materials for use at elevated temperature corrosive environments. The scientific community is poised to utilize these materials specifically in the area of advanced gas turbine engines [3].

The primary constraint precluding the application of CMCs in advanced gas turbine engines is the lack of adequate material characterization and predictive modeling to substantiate sufficient life limiting mechanisms such as creep, fatigue, thermal shock induced rupture, thermal fatigue under constant applied stress, thermomechanical fatigue, corrosion, and oxidation. Most potential CFCC applications involve concurrent thermal and mechanical cycling in hostile environments. In order to generate material data that can be used by the designer, experiments should be conducted under cyclic conditions simulating the component's service environment. However, research data simulating the cyclic service conditions of gas turbine engines such as burner rig thermal fatigue under jet fuel impingement and constant applied stress or better jet fuel burner rig thermomechanical fatigue are not available.

The thermal fatigue study of CFCCs has been limited to a glass ceramic composite, photon heating, and thermal cycling in the temperature range 500–1100°C [4]. Likewise, thermomechanical fatigue research of CFCCs has been limited to glass ceramic composites, photon [5] or induction [6] heating, and thermal cycling between 500–1100°C. Such heating techniques lack the rapid heating and corrosion effects of jet fuel flame impingement. Thus, they are inadequate to simulate the complex interaction of the thermomechanical loading and erosive and corrosive environment of a gas turbine engine. Further, CFCCs in gas turbine applications are expected to operate in a larger cyclic temperature range (500–1350°C).

The burner rig thermal fatigue testing under constant applied stress is nearly the ideal test technique to simulate a jet engine thermal, mechanical, and chemical environment, although in- or out-of-phase concurrent cycling of the mechanical load would be desirable. The advantages of the present test technique over furnace thermal cycling include the impingement of a jet fuel flame at high velocities (0.3 Mach), rapid heating of the specimen, and a constant applied stress. The high-velocity impingement of jet fuel flame also probes material's resistance to erosion under service conditions. The specimen is subjected to thermal shock in both the flame-impinged region and in regions outside the heated zone where high-temperature gradients exist. Thus, failure may occur in either locations. High thermomechanical stresses outside the heated zone may bring about cycle by cycle damage accumulation in these regions.

In the present study, the burner rig thermal fatigue behavior of a CFCC (crossply $HP-SiC_f/Si_3N_4$) under impinged jet fuel flame has been studied for the first time. The specimens were thermal cycled between 500–1350°C under high-velocity (Mach 0.3) jet fuel (JP4) flame impingement and under constant applied stress. This simulates the hot gas path or high-pressure turbine nozzle of a gas turbine engine environment well. The temperature excursions simulate the potential quick temperature "bursts" experienced during vehicle takeoff and rapid cooling during engine shutdown. $HP-SiC_f/Si_3N_4$ is a viable high-temperature material with a potential of surpassing the present operating temperature limit of nickel-based super alloys. An understanding of damage mechanisms and lifetime in thermal fatigue under impinged jet fuel flame and constant applied stress would accelerate its application as a high-temperature engineering material.

EXPERIMENTAL PROCEDURE

MATERIAL AND SPECIMEN GEOMETRY

The 3.175-mm thick 0.30 volume fraction $[(0/90)_3/\overline{0}]_s$ SCS-6® or $[(0/90)_5/\overline{0}]_s$ SCS-9® SiC fiber-reinforced Si_3N_4 crossply laminates were hot pressed at 1650°C under 17.2 MPa pressure for 1 hour by Textron Specialty Materials, Lowell, Massachusetts. The matrix composition was 92.5% Stark LC-12 Si_3N_4, 5.0% Y_2O_3, 1.5% MgO, and 1.0% Al_2O_3, all in weight percentage. Rectangular specimens were machined to 152.4 mm × 12.7 mm × 3.175 mm dimension. This geometry was preferred to round edge geometry because of ease of machining and less likelihood of introducing fabrication defects. The 142-μm diameter SCS-6 and 79-μm diameter SCS-9 SiC fibers, also fabricated by Textron by chemical vapor deposition, have a fine columnar β-SiC emanating radially from a 33-μm diameter carbon monofilament substrate. Only the SCS-6 SiC fibers have a carbon-rich region outside the carbon core. The SCS-9 fiber and the outer zone of the SCS-6 fiber have stoichiometric SiC compositions. Both fibers have approximately 3-μm complex carbon-rich coating at the surface.

TEST APPARATUS

Burner rig thermal fatigue tests under constant applied stress were conducted using the Mach 0.3 atmospheric pressure durability testing burner rig at the NASA Lewis Research Center, Cleveland, Ohio. The specimen temperature was maintained to ±1.7°C and impinging gas velocity to Mach ±.02 [7]. Figure 4.1 shows the flame impingement of HP-(SCS-9)SiC_f/Si_3N_4 in the burner rig test cell. Not shown in the view are the central control console and a personal computer for system control and data acquisition. Figure 4.2 illustrates a close-up view of the flame impingement process. The severe test conditions and specimen failures were recorded using a video recording camera.

FLAME PRODUCTION AND IMPINGEMENT

The flame was produced using a simple "can-type" combustion chamber (Figure 4.3). Preheated air at 260°C and JP-4 fuel with an equivalence ratio of 0.5 was first pumped into the combustion chamber to initiate the combustion process. The flow of the compressed air and fuel were controlled using temperature and pressure sensors in the combustion chamber. In the event of insufficient fuel or feed air flow to the burner rig, an automatic system shutdown would occur.

The geometry of flame impingement onto the specimen is illustrated in Figure 4.4. Approximately 27 mm of the leading edge of the test specimen

Figure 4.1 Flame impingement onto crossply SiC continuous fiber Si_3N_4 ceramic composite specimens in the NASA Lewis Research Center Mach 0.3 atmospheric pressure burner rig test cell.

Figure 4.2 Close-up view of the flame impingement process.

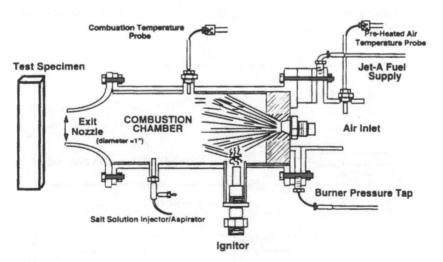

Figure 4.3 Combustion chamber of the NASA Lewis Research Center Mach 0.3 atmospheric pressure burner rig.

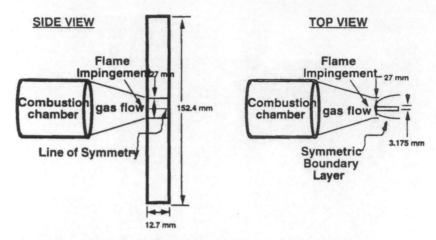

Figure 4.4 Flame impingement configuration showing top and side view of specimen.

was impinged. The specimen was located 25.4 mm away from the exit nozzle. The flame yielded a symmetrical temperature profile above and below the centerline of the specimen.

The three regions of turbulent-free jets of the flame are shown in Figure 4.5. The core region occupies four to five nozzle diameters immediately downstream from the nozzle where the velocity and temperature distribution of the exhaust gas remain constant [8,9]. The specimen is located one nozzle diameter downstream from the nozzle exit where the maximum exit

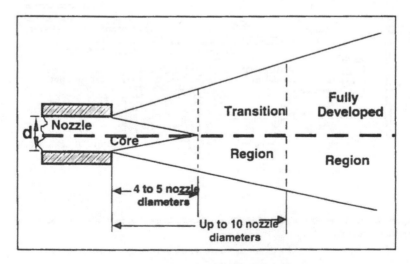

Figure 4.5 Regions of turbulent-free jets.

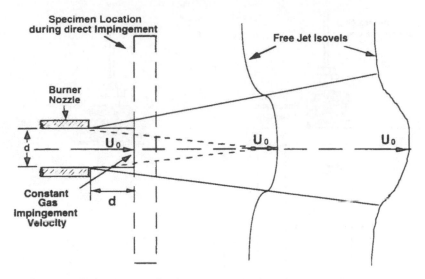

Figure 4.6 Exit velocity profile of free jet showing lines of constant relative velocity.

gas velocity and temperature in both the horizontal and vertical directions are maintained. Beyond the core region, there is a transition zone extending up to 10 nozzle diameters, followed by a fully developed region. The lines of constant relative velocity (isovels) of the existing gas are depicted in Figure 4.6.

LOAD CELL AND GRIPS

A pneumatic load cell actuated by compressed air (Fairlane H4X4HC) was bolted to the loading frame (Figure 4.7). The pneumatic load cell accommodated the cyclic thermal expansion and contraction of the specimen while applying a constant stress up to 710 MPa (500 psig) for the specimen geometry used. The maximum axial displacement of the load cell was 57.2 mm. An end-limit switch activated by specimen failure electronically powered down the burner rig and periphery devices.

The specimens were gripped vertically by end loading using simple collet-type, self-aligning loading grips (ATS 713C). The grips had no active water- or air-cooling circuits. This is an advantage as cooled grips have caused thermal stress concentrations at the grips causing failure in the grip section [10]. No specimen failure in the grip section was observed. Copper bands were wrapped around the specimen in the grip area. There was no indication that slip occurred at the grips.

Specimen alignment was verified using ASTM Standard E1012-89. A

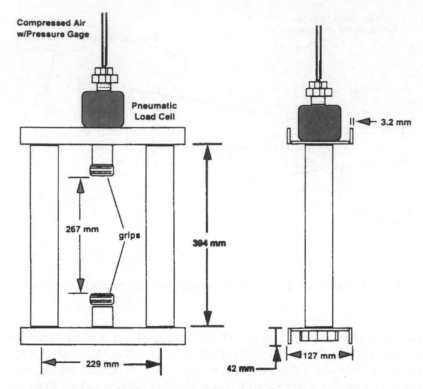

Figure 4.7 Front and side view of loading frame and pneumatic load cell.

25.4-mm gage length extensometer was attached to one side of an aluminum specimen. Longitudinal strain measurements were recorded at approximately 300 MPa. This procedure was repeated for each side of the bar. Bending stresses were found to be less than 1% of the total stress, indicating good load train alignment.

TEMPERATURE CONTROL

A feedback control system using a two-color optical pyrometer was used to control the specimen temperature. The pyrometer was focused on the leading edge of the flame-impinged surface. The flame temperature near the leading edge during steady-state heating was measured with a hand-held R-type ceramic coated thermocouple. No radiation shield was used around the tip of the thermocouple. The measured flame temperatures were within 3% of pyrometer readings of the specimen leading edge. Due to the severe temperature and, what is more important, the high-velocity flame environment, thermocouples and strain gages could not be attached to the specimen to collect experimental data.

A 66°C temperature difference along the specimen width in the flame-impinged zone at 1350°C was measured using the optical pyrometer. The most extreme temperature gradient occurred outside the flame impingement zone. The temperature gradient at the leading edge outside the heated section was estimated to be 529°C (985°F) within approximately 5-mm vertical distance. The trailing edge temperature gradient was 588°C (1091°F) within approximately 5-mm vertical distance. The temperature at the grips was approximately 28°C (83°F).

Although the pyrometer readings were in agreement with thermocouple flame temperature measurements, the location of the measurement along the length or width of the specimen had an uncertainty of approximately ±1 mm. Thus, a finite-element analysis of temperature distributions and accompanying thermal stresses during both steady-state and transient thermal loading was carried out, which will be the subject of a separate publication [11].

TEST MATRIX

An experimental test matrix for thirteen test specimens was developed, incorporating the capabilities of the test system. A "set-up" test specimen was used to evaluate the transient temperature profile during heating and cooling. The time to run out was arbitrarily selected as 1000 cycles due to the cost of operating the test facility. The applied stress was varied from specimen to specimen to reflect the learning of the composite system behavior under the severe thermal cycling. Constant tensile stresses 16.8, 84, 110, 120, 125, 130, 140 and 168 MPa were applied to individual specimens during thermal cycling.

THERMAL CYCLING

Thermal cycling of the specimens between 500–1350°C was accomplished by a computer-controlled hydraulic piston (shown in Figure 4.1) that rotated the burner rig nozzle into and out of the direction of the specimen. An electronic timing device controlled the duration of the thermal cycle, and a cycle time recorded the number of heating and cooling half cycles to failure.

The length of the heating half cycle (60 s) was dictated by the test system's capabilities. The combustion chamber temperature and pressure control system required approximately 50 s to adjust process parameters and obtain feedback from the system's components. This required the heating half cycle to be at least 60 s. Incorporating a 60-s heating half cycle eliminated specimen temperature overshoot as well as minimized "under-temperature" conditions during start-up. The test facility required between three to five heating half cycles to achieve desired temperature. The setup

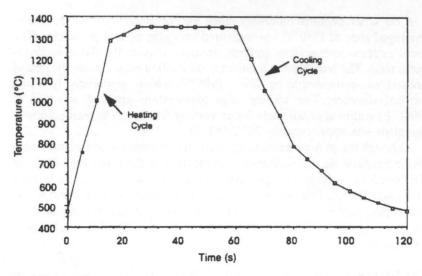

Figure 4.8 Transient specimen leading edge temperature profile during burner rig thermal cycling showing the heating and cooling half cycles.

specimen required approximately 60 s to cool to 500°C by radiation and natural convection after a 60-s heating half cycle. A symmetrical time parameter of 60 s was selected for cooling.

The specimen leading edge cyclic temperature profile resulting from a 60-s 1350°C flame impingement, followed by a 60-s passive, ambient temperature cooling is shown in Figure 4.8. The temperature of the impinging gas was a function of the combustion fuel, feed flow of compressed air and burner rig pressure. The maintained exiting gas temperature of 1350°C is near the peak temperature capability of the facility.

POST-TEST EVALUATION

Post-test evaluation of test specimens involved fractographic examination of failed specimens using scanning electron microscopy, room temperature tension tests on runout SCS-9 SiC fiber-reinforced specimens for residual strength, and SIMS (secondary ion mass spectroscopy) and XPS (X-ray photoelectron spectroscopy) surface analyses of both the flame-impinged zone and the unheated areas at the end of the sample for reference. The tension tests were conducted before and after thermal cycling under load control using a servohydraulic universal testing machine. Composite tabs were attached to the rectangular specimens using epoxy adhesive. Tension tests for residual strength on the runout SCS-9 continuous fiber specimens provided information on any cycle-by-cycle mechanical damage accumulation.

RESULTS

BURNER RIG THERMAL FATIGUE UNDER CONSTANT APPLIED STRESS

The burner rig thermal fatigue lifetimes of nine $[(0/90)_s/\bar{0}]_s$ SCS-9 and four $[(0/90)_3/\bar{0}]_s$ SCS-6 SiC continuous fiber specimens under constant applied stress are given in Figure 4.9. The SCS-9 specimens showed superior burner rig thermal fatigue resistance at any constant applied stress (Figure 4.9). The SCS-6 specimens failed after twenty to forty cycles at constant applied stresses as low as 110 MPa. The SCS-9 specimens, on the other hand, showed run out conditions (1000 cycles without failure) up to 125 MPa constant applied stress. The number of cycles-to-failure in the SCS-9 continuous fiber specimens declined abruptly to 115 thermal cycles at 140 MPa constant applied stress. At 168 MPa, applied stress fatigue life diminished to five to forty-two cycles (Figure 4.9).

RESIDUAL STRENGTH

Two runout SCS-9 specimens thermal cycled under 84 and 105 MPa constant applied stress were tension tested at room temperature. The

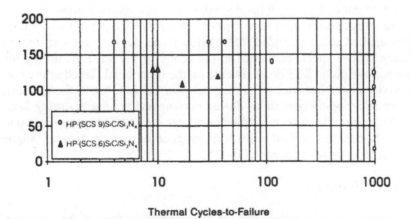

Thermal Cycles-to-Failure

Figure 4.9 Burner rig thermal fatigue response of HP-(SCS-6)SiC$_f$/Si₃N₄ and HP-(SCS-9) · SiC$_f$/Si₃N₄ crossply ceramic composites under constant applied stress.

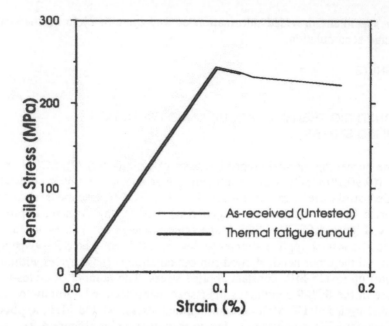

Figure 4.10 Room temperature tensile stress-strain curves of as-received and burner rig thermal fatigue runout HP-(SCS-9)SiC$_f$/Si$_3$N$_4$ composite specimens.

stress-strain curves of an untested (control) sample and burner rig runout specimens are shown in Figure 4.10. The slope in Figure 4.10 signifies a modulus of 260 GPa (37.8 Msi). The tensile strength of the as-received and thermal fatigue runout specimens was found to be 244 MPa. These indicate no strength or modulus degradation in the runout samples.

The video recording of each burner rig thermal fatigue test under constant applied stress provided information about the location and time of failure. SCS-9 specimens failed in the center of the hot zone during the heating half cycle. SCS-6 specimens, on the other hand, failed at the bottom edge of the hot zone where the temperature gradient is the greatest. Three SCS-6 specimens failed during cooling and the fourth during heating half cycle. The runout conditions exhibited by SCS-9 specimens up to 125 MPa constant applied stress indicate good burner rig thermal fatigue resistance in cycles between 500–1350°C.

FRACTOGRAPHY

Scanning electron microscopy of the fracture surface of the specimen that survived 1000 cycles under 84 MPa applied stress and subsequently tension tested at room temperature showed fiber pullout less than 1 mm

(Figures 4.11 and 4.12). No interface oxidation is indicated. Specimens that failed during thermal cycling showed similar fracture features; in some, 33-μm diameter carbon core was pulled out (dangling) in unoxidized condition.

SURFACE ANALYSIS

Changes in surface chemical composition of the flame-impinged zone, relative to the grip areas, were examined using SIMS (secondary ion mass spectroscopy) and XPS (X-ray photo electron spectroscopy) techniques. The grip areas were also examined for reference. The sintering aid Y (Y_2O_3) showed a significant relative decrease in surface concentration in the grain boundary glassy phase in the flame impinged zone. The concentrations of alkali metals Na and K showed increased relative to the reference grip areas. The Si concentration diminished by about four times with respect to Al in the flame-impinged zone. The runout specimens examined

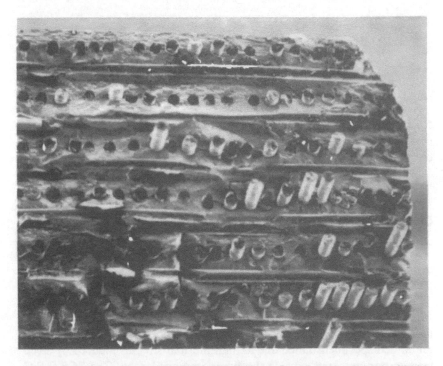

Figure 4.11 SEM fractograph of HP-(SCS-9)SiC$_f$/Si$_3$N$_4$ specimen thermal cycled under 85 MPa constant applied stress and survived 1000 cycles (runout), tension tested at room temperature (45×).

Figure 4.12 Close-up of SEM fractograph of the specimen shown in Figure 4.10 (300×).

for surface cracks at 200× magnification using scanning electron microscopy showed no surface cracks. Thus, the chemical changes in the surface
did not lead to the production of surface cracks.

DISCUSSION

The thirteen-ply SCS-6 and twenty-one–ply SCS-9 SiC fiber 0/90 Si_3N_4
ceramic composites thermal cycled between 500–1350°C under impinged
jet fuel flame and a constant applied stress showed a different burner rig
thermal fatigue behavior. SCS-9 SiC continuous fiber specimens showed
superior burner rig thermal fatigue resistance than SCS-6 SiC fiber-
reinforced Si_3N_4 at any constant applied stress. The SCS-6 SiC fiber-
reinforced specimens failed due to high thermal stresses induced by steep
temperature gradients outside the flame-impinged zone. The SCS-9 SiC
fiber-reinforced specimens, however, survived these high thermal stresses
and showed burner rig thermal fatigue runout conditions (1000 cycles

without failure) at applied stresses greater than 50 percent of the room temperature strength.

Young's modulus of the SCS-6 fiber is 390 GPa, and the Young's modulus of the SCS-9 fiber is 331 GPa. This is because the volume fraction of the low modulus carbon core is SCS-6 and SCS-9 SIC fibers are 5.4 and 17.5 percent, respectively. The Young's modulus of the SCS-9 SiC fiber-reinforced specimens was measured to be 260 GPa at room temperature. Using the laminated plate theory, this gives, for the SCS-6 SiC fiber-reinforced specimens, a Young's modulus of approximately 280 GPa. Assuming that both SCS-6 and SCS-9 specimens have the same coefficient of thermal expansion and were subjected to the same temperature gradients, the thermal stresses outside the flame-impinged zone in the SCS-6 specimens would be 8% higher than that in the SCS-9 specimens. It appears that this increased stress in the SCS-6 specimens induced accumulation of cycle-by-cycle damage outside the heated zone leading to early failure in these locations.

If the failure occurs in the heated zone, burner rig thermal fatigue testing under constant applied stress crosses the boundary between elevated temperature tensile fatigue and creep testing. The tension tests on the SCS-9 runout specimens showed no degradation of modulus or strength. Thus, the failure of the SCS-9 specimens could not occur due to a cycle-by-cycle accumulation of thermal fatigue damage in the flame-impinged zone; however, creep or thermal instability of the SCS-9 fibers in the heated zone could bring about failure of the specimen. The abrupt decline in the number of cycles to failure in SCS-9 ceramic composites in the constant applied stress range 140–168 MPa and lack of thermomechanical fatigue damage indicate tensile creep failure of the composite.

Whether the transverse layers in a crossply composite will contribute to the creep resistance of the composite depends on the creep behavior of the matrix. If the matrix creep strain rate is higher than that of the fiber, a transfer of stress from the matrix to the fibers takes place as a result of matrix relaxation. In that case, the transverse layers will not contribute to the creep resistance of the 0/90 composite significantly. In hot pressed SiC/Si₃N₄ composites, the matrix has poor resistance to creep due to the grain boundary glassy network [12], whereas the fibers exhibit good resistance to creep [13].

As mentioned earlier, the measurement of strain while the flame is impinged is difficult. In the absence of accumulating strain data, a concentric cylinder, micromechanical finite-element model could be developed to estimate the creep rate and cycle-by-cycle strain ratchetting. This could be done for a hold time of approximately 45 s per cycle at 1350°C and approximately 5 s for heating and cooling half cycles. However, in a recent study, Butkus, Holmes, and Nicolas [5] reported that the strain accumulation in thermomechanical fatigue of a SiC fiber-reinforced calcium aluminosilicate composite was faster than that in isothermal fatigue under the same loading conditions. Thus, the micromechanical model could not account for strain enhancement due to cycling and would underestimate the evolution of strain in the composite. Further, Xin and Holmes [14] found equal creep rates in unidirectional and crossply SiC fiber-reinforced calcium aluminosilicate ceramic composites, indicating that the creep compliant matrix might have contributed to the creep resistance of the composite. Thus, a concentric cylinder micromechanical model could not adequately describe the evolution of strain in burner rig thermal fatigue.

Bhatt and Hull [15] studied the thermal instability and accompanying strength degradation in three chemically vapor deposited Textron SiC fibers after exposure to 1400°C under argon atmosphere for 1 h. They found that strength degradation occurred due to recrystallization and grain growth of the near-stoichiometric SiC grains near the surface of the fibers. They attributed this strength degradation to the impurities such as C and Si present at the grain boundaries. In the present case, the SCS-9 fibers were exposed to 1350°C for more than 6 h under constant applied stresses up to 125 MPa in the runout specimens. During this time of exposure and under stresses above 50 percent of the room temperature strength of the composite, the fibers are likely to undergo strength degradation by recrystallization grain growth. Further, the stresses in the 0° plies would be above that of the applied stress. Using the laminated plate theory, the stress in the 0° layers for an applied stress of 168 MPa is estimated to be 198 MPa. Under these conditions, the most likely mode of failure for the SCS-9 SiC fiber-reinforced specimens would be the thermal instability induced failure of fibers.

CONCLUSIONS

The SCS-9 continuous fiber runout specimens thermal cycled under constant applied stresses up to 105 MPa showed no significant strength or modulus degradation. Likewise, the fractographic examination of thermally cycled and failed SCS-9 specimens showed no internal oxidation. Minor changes in the surface chemistry did not induce cracks or roughness in the surface. Thus, the failure of SCS-9 SiC fiber-reinforced specimens could not have taken place by cycle-by-cycle thermal fatigue damage accumulation. The abrupt decline in the burner rig fatigue life of the SCS-9 specimens in the constant applied stress range 140–168 MPa, coupled with lack of microstructural damage and time of exposure at 1350°C, indicate the likelihood of thermal instability induced rupture of the fibers. The failure of the SCS-6 specimens is most likely due to cycle-by-cycle damage induced by high thermomechanical stresses outside the flame-impinged zone.

ACKNOWLEDGEMENTS

The author is grateful to NASA Lewis Research Center, Cleveland, OH, for the permission to use their burner rig test station. The author would like to express his appreciation to Dr. George St. Hilaire for useful discussions and arranging the collaboration between University of Massachusetts and NASA Lewis Research Center.

REFERENCES

1 Evans, A. and Marshall, D. "The Mechanical Behavior of Ceramic Matrix Composites," *Acta Metall.*, Vol. 37, 1989, p. 2583.

2 Argon, A. S. and Gupta, V. "Control of Toughness of Composites through Control of Strength of Fiber/Matrix Interfaces," *Damage and Oxidation of Protection in High Temperature Composites*, G. K. Haritos and O. O. Ochoa, eds., ASME Book No. H0692B, 1991.

3 Kannmacher, K. and Groseclose, L. "Design and Analysis of Ceramic and CMC Components for Advanced Gas Turbines," presented at the *International Gas Turbine and Aeroengine Congress and Exposition*, Orlando, FL, June 3–6, 1991, ASME 91-GT-156.

4 Zawada, L. and Wetherhold, R. "The Effects of Thermal Fatigue on a SiC Fibre/Aluminosilicate Glass Composite," *J. Mater. Sci.*, Vol. 26, 1991, p. 654.

5 Butkus, L. M., Holmes, J. W., and Nicholas, T. "Thermomechanical Fatigue of a Silicon Carbide Fiber-Reinforced Aluminosilicate Composite," *J. Am. Ceram. Soc.*, Vol. 76, 1993, p. 2817.

6 Worthem, D. W. and Ellis, J. R. "Thermomechanical Fatigue of Nicalon CAS [0]₃ₛ under In-Phase and Out-of-Phase Cyclic Loadings," *Ceram. Eng. Sci. Proc.*, Vol. 14, No. 7–8, 1993, pp. 292–300.

7 Deadmore, D. L. "Digital Temperature and Velocity Control of Mach 0.3 Atmospheric Pressure Durability Testing Burner Rigs in Long Time, Unattended Cyclic Testing," NASA Technical Memorandum 86959, March 1985.

8 Beer, J. M. and Chigier, N. A. *Combustion Aerodynamics*. Robert E. Kreiger Publishing Company, 1972.

9 Lefebrve, A. H. *Gas Turbine Combustion*, McGraw-Hill, 1983.

10 Holmes, J. W. "A Technique for Tensile Fatigue and Creep Testing of Fiber-Reinforced Ceramics," *J. Compos. Mater.*, Vol. 26, 1992, p. 932.

11 Ertürk, T. "Heat Transfer during Burner Rig Thermal Cycling of SiC Fiber Silicon Nitride Ceramic Composites: A Finite Element Analysis," to be published.

12 Kossowsky, R., Miller, D. G. and Munz, D. "Tensile and Creep Strengths of Hot Pressed SiC$_f$/Si₃N₄," *J. Mater. Sci.*, Vol. 10, 1975, pp. 983–997.

13 DiCarlo, J. A. "Creep of Chemically Vapour Deposited SiC Fibers," *J. Mater. Sci.*, Vol. 21, 1986, pp. 217–224.

14 Xin, W. and Holmes, J. W. "Tensile Creep and Creep-Strain Recovery Behavior of Silicon Carbide Fiber/Calcium Aluminosilicate Matrix Ceramic Composites," *J. Am. Ceram. Soc.*, Vol. 76, 1993, pp. 2695–2700.

15 Bhatt, R. T. and Hull, D. R. "Microstructural and Strength Stability of CVD SiC Fibers in Argon Environments," *Ceram. Eng. Sci. Proc.*, Vol. 12, No. 9–10, pp. 1832–1844.

CHAPTER 5

High-Temperature Compression Test Technique for Continuous Fiber Ceramic Composites

TURGAY ERTURK[1] and PETER D. MILLER[2]

OVERVIEW

A high-temperature compression test apparatus for continuous fiber ceramic composites is described. Test methodology and some experimental results on the rate independent strength and fracture behavior of a hot pressed crossply and unidirectional silicon carbide fiber/silicon nitride matrix (SiC_f/Si_3N_4) ceramic composite at both room temperature and 1400°C are presented.

INTRODUCTION

Although the high-temperature tensile strength, fatigue, and creep behavior of continuous fiber ceramic composites (CFCC) have been studied extensively [1–8], little work has been reported on their compressive behavior [9–12]. This is presumably due to their high expected strength in compression, as is the case in monolithic ceramics. However, in some ceramic composite systems, the compressive strength has been reported to be as low as the tensile strength. Lankford [9] reported an increase in compressive strength with increasing temperature in SiC_f/LAS glass matrix and SiC_f/SiC composites. Further, the fracture behavior in compression of CFFCs may differ from that of other continuous fiber composites. Thus,

[1]University of Massachusetts Lowell, Department of Chemical and Nuclear Engineering, Lowell, MA, U.S.A.
[2]Purdue University, Department of Materials Engineering, W. Lafayette, IN, U.S.A.

an examination of the compressive behavior of brittle matrix composites is pertinent.

Continuous fiber composites often fail in compression at low temperatures by the buckling of fibers supported by a compliant matrix leading to the formation of kinks in the specimen. Although the fiber elastic instability mode of failure in compression has been well documented [13–17], compressive failures by vertical splitting or slanted shear faulting of the composite specimen have also been reported [9–11,18–23]. The room temperature compressive behavior of monolithic brittle solids has also been well characterized [24–27]. The dominant mechanisms of failure, splitting and shear faulting, result from the formation of columns in the material as a result of interaction of secondary wing cracks.

It can be expected that, as the modulus of the matrix approaches that of the fiber in a continuous fiber brittle composite, the failure mechanism will shift from fiber buckling to specimen splitting and shear fault formation. The interaction of matrix secondary wing crack tip stresses with each other and with residually stressed fiber interfaces [28] would play a key role in controlling the mode of compressive failure. The failure mechanisms would remain essentially the same at room temperature and at elevated temperatures with no rate effects. Low toughness and favorably residually stressed fiber interfaces in a matrix composite may provide easy paths for crack propagation along interfaces, leading to compressive weakening of the composite. The failure mechanisms at high-temperature creep conditions would differ significantly.

In the present study, a high-temperature compression test apparatus for contiuous fiber ceramic composites is described, and some experimental results on the compressive strength of Textron Specialty Materials' SCS-6 SiC fiber-reinforced crossply and unidirectional SiC_f/Si_3N_4 ceramic composite at room temperature and 1400°C are reported.

HIGH-TEMPERATURE COMPRESSION TEST APPARATUS

An IITRI (Illinois Institute of Technology Research Institute)-type compression test apparatus was built using a preexisting plastic injection mold base. The Celanese and IITRI compression test techniques were originally developed for polymer composites. Both test techniques use plate specimens side gripped at the ends. They have been standardized for room temperature measurements (ASTM D3410). The Celanese compression test fixture has a circular geometry, as shown in Figure 5.1, whereas the IITRI apparatus utilizes a rectangular geometry.

The IITRI technique provides a larger space for the local heating of the specimen gage length. Another advantage of the IITRI type of text fixture

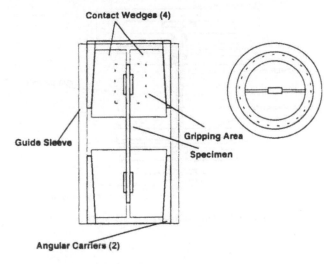

Figure 5.1 Celanese-type compression test apparatus.

over the Celanese type is the construction cost. The Celanese fixture requires high-cost machining to obtain the needed tolerances in mating conical components. Also, the cooling of the compression apparatus is easier in the IITRI method due to the presence of large available space.

A schematic cross-sectional view of the modified ITTRI-type, high-temperature compression test apparatus is shown in Figure 5.2. A heating

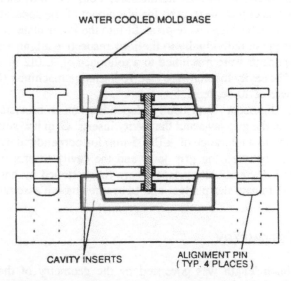

Figure 5.2 Modified IITRI-type, high-temperature compression test fixture.

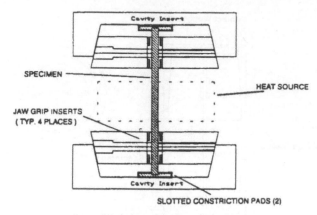

Figure 5.3 Grip geometry and alignment.

oven was placed between the upper and lower dies. The apparatus was designed to operate at specimen temperatures up to 1400°C and grip temperatures up to 800°C.

The grip/jaw area of the apparatus (shown in Figure 5.3) receives the greatest heat flux from the heated zone. The grip jaws and cavity inserts in contact with the specimen were made of an air-hardening medium alloy tool steel (A-2) hardened to 56 Rockwell C to resist temperatures up to 800°C and galling wear.

A pocket was machined in each grip jaw to accommodate interchangeable friction inserts of various hardnesses. Four 19.1-mm diameter case hardened tool steel pins were used for alignment of the apparatus (Figure 5.2). Four bronze bushings were press-fit into the lower plate of the fixture and then honed out individually to the four respective alignment pins. All mating components were machined to a tolerance of ±.012 mm to ensure alignment. The cavity inserts were electrodischarge machined (EDM) to a reference angle of 10 degrees.

Specimen alignment is critically dependent on a precision angular match between the grip jaws and the cavity inserts. Grip jaw surfaces were surface ground to a tolerance of ±0.005 mm for perpendicularity, and the angular match between the grip jaws and the cavity inserts was held to within ±0.016 degrees. All surfaces were finish stoned by hand prior to assembly to eliminate sharp edges and to ensure smooth assembly and operation.

SPECIMEN GEOMETRY

The specimen length was governed by the geometry of the available CFCC test panels having a dimension 115 × 115 × 5 mm. Specimen gage length was also limited by concern over Euler buckling. The uncontoured test samples were 115 mm long, 12.7 mm wide, and 3.2 mm thick. A gage length of 60 mm was exposed to oven heating, leaving approximately 27.5 mm grip length at each end. This proved insufficient in generating adequate frictional traction at the grips to load the specimen; the specimen slid through the grips and contacted the cavity insert. This produced brooming of the specimen ends. Slotted pads were mounted to the cavity inserts (Figure 5.3). This eliminated the brooming problem [9].

Specimens were machined laterally to 38.1-mm diameter circular contour, as shown in Figure 5.4, giving a 3.2 × 3.2 mm square cross-sectional area. The square gage cross section was preferred because of the possibility of differences in fiber packing arrangement in the transverse direction. The reduced cross section gage geometry is similar to the specimens successfully used by Larsen and Stuchly [12]. The circular contour was used primarily due to the difficulty in machining complex geometries. It is recognized that preferential geometries such as the "streamline" specimens developed by Oplinger et al. [29,30] greatly reduce the shear stresses within the stress gradient regions. However, computer numeric controlled diamond tool grinding would be required to machine such geometries. The stress concentration due to the gage reduction was determined using the finite-element technique using SDRC-Ideas. The finite-element mesh used is shown in Figure 5.5. A stress concentration of 2.7 percent in the axial direction, and 0.38 percent in the transverse direction was found. The finite-element model of the specimen converged to within 2 percent. More complex geometries could equally be used to reduce gage stress concentrations.

The contoured specimen geometry forced the fracture to occur in the reduced section, facilitating the in situ optical observation (and recording) of the failure process at the ambient temperature. It also secured temperature uniformity in the fracture zone at elevated temperatures. Further, the localization of the fracture process eliminated any rate effects. If the matrix attained rate-dependent behavior at elevated temperatures, there would be a higher probability of failure in the cooler regions. The major objection to utilizing this geometry is that of forcing the failure to a small volume.

HEATING OVEN

Silicon carbide resistance heaters were used by recommendation of Starrett [31]. The oven consisted of two matching halves (Figure 5.6), each equipped with two Norton silicon carbide resistance heating elements ("igniters") bonded to the oven using Autocrete, a high-temperature, ceramic adhesive from Flexbar Machine Corporation. Norton's Alundum bubble alumina, having an application temperature of 1870°C, was used for the

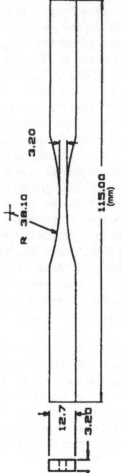

Figure 5.4 Specimen geometry having a square cross-sectional area in the center.

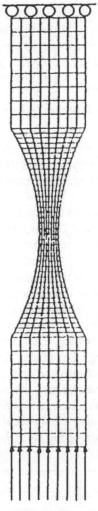

Figure 5.5 Finite-element model of specimen.

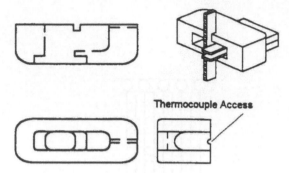

Figure 5.6 High-temperature oven construction.

oven material. The oven was machined to a cavity size of 19 × 51 × 28 mm using conventional masonry tools. This design was found to be cost-effective and simple to manufacture. A schematic end view of the heating system is shown in Figure 5.7.

Preliminary experiments indicated that radiation shields were necessary to maintain the temperature of the IITRI fixture at a level less than 800°C. Successful radiation shields were made of 3.2-mm thick Fiberfrax mat (available from Carborundum Inc.) placed between two 0.05-mm thick 304 stainless steel sheets, as shown in Figure 5.8. The radiation shields were held together using 0.25-mm diameter Monel wires.

The silicon carbide resistance heating elements used operate at voltages up to conventional 110 line voltage, eliminating the need for the use of complex power supplies. The heating elements are stable in air for the period of time over which testing takes place. The heating elements are inexpensive. This is important because the specimen shatters upon failure and can damage heaters and oven. From the experimentally determined optimal spacing of these heaters, the maximum temperature variation at the reduced section of the specimen was approximately 30°C at a nominal temperature of 1400°C, as shown in Figure 5.9. A digital multimeter was

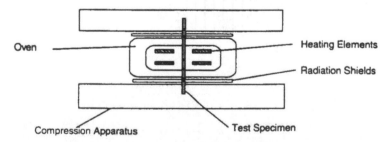

Figure 5.7 Schematic end view of apparatus demonstrating placement of oven parts.

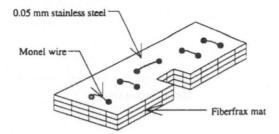

Figure 5.8 High-temperature radiation shields.

used to measure output voltage of the three constant temperature ice bath referenced R-type thermocouples.

The temperature distribution of the test apparatus (across the oven) was determined experimentally. Six independent tests were conducted with a step input of 200 W per heater. The temperature at the specimen center was monitored using only one R-type thermocouple. Other temperature measurements were made using K-type thermocouples. These results are shown in Figure 5.10.

The power input to each of the four heating elements was monitored manually using four variable controllers, as shown in Figure 5.11. With a step input of 200 W per heater, the thermal time constant of the system was determined to be less than five minutes. Thus, the time required to stabilize the testing temperature is approximately fifteen minutes. Although manual power control was used in tests, various feedback control schemes could readily be implemented.

EXPERIMENTS

Composite specimens were fabricated using Textron Specialty Materials' 142-μm diameter SCS-6 silicon carbide (SiC) fibers and as-

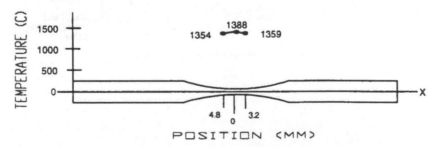

Figure 5.9 Temperature distribution in the reduced section of the specimen.

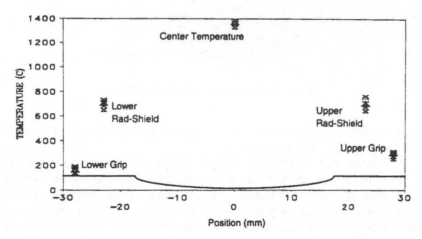

Figure 5.10 Temperature distribution across specimen at various points within oven.

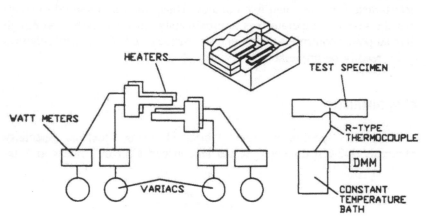

Figure 5.11 Temperature monitoring system.

received SN-E10 silicon nitride (Si_3N_4) powder manufactured by UBE Industries. The matrix composition was 5.0% Y_2O_3, 1.5% MgO, 1.0% Al_2O_3, and the remainder Si_3N_4. Powders were ball milled separately. A water-soluble acrylic binder was added to the slurry for preform handleability. The fibers were wound on a drum attached to a lathe at a spacing of forty-one fibers per centimeter, using a specially designed cross-feed fixture. The powder mixtures were then sprayed onto the aligned fibers with an air-brush kit to a thickness of 0.5 mm and allowed to dry. The preform was removed by cutting along a machined groove in the winding drum and further cut to the desired panel size.

The layup was hot pressed for 1 h at 1650°C under a pressure of 17.2 MPa at Textron Specialty Materials, Lowell, Massachusetts. The unidirectionally reinforced panel was 114 × 127 × 3.2 mm (machined to eight plies), and the 0/90 crossply panel was 114 × 127 × 4.8 mm (machined to thirteen plies). Panels were machined to the hourglass geometry (Figure 5.4).

Compression test results of SiC_f/Si_3N_4 composites at room temperature and at 1400°C are given in Table 5.1. The results indicate relatively high compressive strengths at both temperatures. For unidirectionally reinforced specimens, the strength decreased by approximately 30 percent between ambient temperature and 1400°C. This is in variance with observations of Lankford [9] for a lithium aluminosilicate (LAS) glass-ceramic reinforced with Nicalon SiC fibers. In glass-ceramic composites, crack healing and interface carbonization effects may enhance the strength at elevated temperatures. The scatter in the ambient temperature strength of the unidirectional and elevated temperature strength of the crossply specimens are high (Table 5.1). The scatter in the elevated temperature test results of the unidirectionally reinforced specimens, on the other hand, is only 1.3 percent.

A Sony CCD color video camera with an adapter for a 6× objective was used for real-time video observation and recording of ambient temperature tests and post-test analysis of surface damage. An approximately 20-μm length on the surface of the specimen could be magnified to 40 cm on a video screen; however, the clarity of the images was seriously impaired by extraneous building and machine vibrations and the difficulty of adjusting the observation stage with respect to the composite surface. The failure process in room temperature testing of the unidirectionally reinforced composites could not be video recorded because of the explosive nature of the failure process. The failure occurred unstably without any noticeable precracking activity in the matrix.

The room temperature compressive strength test results for the crossply composites are given in Table 5.1. The crossply composites exhibited higher room temperature compressive strengths compared to unidirec-

TABLE 5.1. Compression Test Results of SiC$_f$/Si$_3$N$_4$ Composites.

Layup	Temperature (°C)	Strength (MPa)	Mean (MPa)	Std. Dev. (MPa)
[0]$_6$	25	1241		
[0]$_6$	25	861	1199	319
[0]$_6$	25	1495		
[0]$_6$	1400	851		
[0]$_6$	1400	830	839	11
[0]$_6$	1400	836		
[(0/90)$_3$/$\overline{90}$]$_s$	25	1670		
[(0/90)$_3$/$\overline{90}$]$_s$	25	2050		
[(0/90)$_3$/$\overline{90}$]$_s$	25	1900	1795	221
[(0/90)$_3$/$\overline{90}$]$_s$	25	1560		

tionally reinforced composites. In agreement with Lankford's observations [11], stable vertical splitting in the 90° layers was observed prior to the onset of unstable fracture in recorded video images. The fracture path in 0/90 crossply composite specimens is shown schematically in Figure 5.12. The cracks initiated from transverse fibers due to stress concentrations around them and propagated stably in the vertical direction during approximately 15 s prior to the onset of unstable fracture. During this process, delamination of outer matrix layers took place. Such failures also occur in brittle materials containing pores.

SUMMARY AND CONCLUSIONS

A test apparatus was developed for the high-temperature compression testing of ceramic composites. The modified IITRI compression test apparatus side loads the 12.7 × 12.7 mm × 120 mm specimens. Longer or wider, or both, specimens could also be tested. An alumina oven was used to heat the gage area of specimens at temperatures up to 1400°C in air. Silicon carbide resistance heaters capable of operating from a 110 AC volt power source were used. The temperature variation in the gage length was measured to be approximately 30°C at full operating temperature. The ambient temperature fracture behavior of crossply SiC$_f$/Si$_3$N$_4$ composite specimens could be videotaped using microscopy instrumentation.

The modified IITRI high-temperature compression test apparatus can be used to study the high-temperature creep and fatigue behavior of ceramic composites. The heating elements do not degrade over approximately 100 h exposure at 1400°C in air. High-temperature extensometer rods could be inserted into the oven through the walls containing the heaters.

Because the compressive fracture of unidirectionally reinforced SCS-6 SiC_f/Si_3N_4 composites occurred unstably without observable prior matrix cracking, the failure mechanisms in this material could not be determined accurately. However, it may be suspected that fracture could have initiated at favorably oriented matrix discontinuities (cracks), then propagated vertically, interacting with fiber residual stresses.

The evolution of fracture in 0/90 SCS-6 SiC_f/Si_3N_4 ceramic composites could be videotaped. Cracks initiated at the transverse fiber interfaces and grew vertically between 90° fibers in each ply, creating vertical columns in the material. This was followed by delamination of the outer matrix layers. It is believed that the final failure of these vertical columns occurred catastrophically by buckling. The fracture process was semistable; it took approximately 15 s from the first observation of damage to the final unstable failure. Since failure initiated at 90° fiber interfaces, it can be expected that residual compressive interfacial hoop stresses could enhance the composite compressive fracture strength. This may be the opposite interfacial condition desirable for unidirectionally reinforced brittle matrix composites.

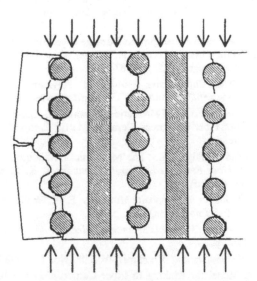

Figure 5.12 Video recorded images clearly showed cracks in crossply composites initiating from transverse fibers and linking vertically. Outer plies were then observed to buckle stably, leading to catastrophic failure.

REFERENCES

1 Bhatt, R. T. "Mechanical Properties of SiC Fiber-Reinforced Reaction-Bonded Si_3N_4 Composites," in *Tailoring Multiphase and Composite Ceramics*, R. E. Tressler, ed., Plenum Pub., 1986.

2 Evans, A. G. and Marshall, D. B. "The Mechanical Behavior of Ceramic Matrix Composites," *Fiber Reinforced Ceramic Composites*, K. S. Mazdiyasni, ed., Noyes, 1990, pp. 1-39.

3 Kodama, H., Sakamoto, H., and Miyoshi, T. "Silicon Carbide Monofilament-Reinforced Silicon Nitride or Silicon Carbide Matrix Composites," *J. Am. Ceram. Soc.*, Vol. 72, No. 4, 1989, pp. 551-558.

4 Holmes, J. W. "Influence of Stress Ratio on the Elevated Temperature Fatigue of a SiC Fiber-Reinforced Si_3N_4 Composite," *J. Am. Ceram. Soc.*, Vol. 74, No. 7, 1991, pp. 1639-1645.

5 Holmes, J. W., Park, Y. H., and Jones, J. W. "Tensile Creep and Creep-Recovery Behavior of a SiC-Fiber-Si_3N_4-Matrix Composite," *J. Am. Ceram. Soc.*, Vol. 76, No. 5, 1993, pp. 1281-1293.

6 Kerans, R. J., Hay, S. H., Pagano, N. J., and Parthasarthy, T. A. "The Role of the Fiber-Matrix Interface in Ceramic Composites," *Ceram. Bull.*, Vol. 68, No. 2, 1989, pp. 429-442.

7 Evans, A. G., He, M. Y., and Hutchinson, J. W. "Interface Debonding and Fiber Cracking in Brittle Matrix Composites," *J. Am. Ceram. Soc.*, Vol. 72, No. 12, 1989, pp. 2300-2303.

8 Cao, H. C. and Thouless, M. D. "Tensile Tests of Ceramic-Matrix Composites: Theory and Experiment," *J. Am. Ceram. Soc.*, Vol. 73, No. 7, 1990, pp. 2091-2094.

9 Lankford, J. "Compressive Strength and Damage Mechanism in a SiC-Fiber Reinforced Glass-Ceramic Matrix Composite," *5th Int. Conf. on Comp. Matls.*, sponsored by TMS, San Diego, 1985, pp. 587-602.

10 Lankford, J. "The Compressive Failure of Polymeric Composites under Hydrostatic Confinement," submitted to *Composites*, in review.

11 Lankford, J. "The Effect of Hydrostatic Pressure and Loading Rate on Compressive Failure of Fiber-Reinforced Ceramic Matrix Composites," *Ceram. Eng. Sci. Proc.*, In Press.

12 Larsen, D. C. and Stuchly, S. L. "The Mechanical Evaluation of Ceramic Fiber Composites," in *Fiber Reinforced Ceramic Composites*, K. S. Mazdiyasni, ed., Noyes, 1990, pp. 182-219.

13 Rosen, B. W. *Mechanics of Composite Materials*. F. W. Wendt, H. Liebowitz, and N. Perone, eds., Pergamon Press, 1970, p. 621.

14 Argon, A. S. *Treatise on Materials Science and Technology*. H. Herman, ed., Academic Press, 1972, p. 79.

15 Budiansky, B. *Computers and Structures*, Vol. 16, No. 1-4, 1983, p. 3.

16 Steif, P. S. "A Model for Kinking in Fiber Composites—I. Fiber Breakage via Micro-Buckling," *Int. J. Solids Structures*, Vol. 26, No. 5-6, 1990, pp. 549-561.

17 Evans, A. G. and Adler, W. F. *Acta Metall.*, Vol. 26, 1978, p. 725.

18 Broutman, L. J., "Failure Mechanisms for Filament Reinforced Plastics," *Modern Plastics*, April 1965, pp. 143-145+.

19 Ewins, P. D. and Potter, R. T. "Some Observations on the Nature of Fibre Reinforced Plastics and the Implications for Structural Design," *Phil. Trans. R. Soc. Lond.*, Vol. A 294, 1980, pp. 507–517.

20 Parry, T. V. and Wronski, A. S. "Kinking and Tensile Compressive and Interlaminar Shear Failure in Carbon-Fibre-Reinforced Plastic Beams Tested in Flexure," *J. Mater. Sci.*, Vol. 16, 1981, pp. 439–450.

21 Bazhenov, S. L., Kozey, V. V., and Berlin, A. A. "Compression Fracture of Organic Fibre Reinforced Plastics," *J. Mater. Sci.*, Vol. 24, 1989, pp. 4509–4515.

22 Hancox, N. L. "The Compression Strength of Unidirectional Carbon Fibre Reinforced Plastic," *J. Mater. Sci.*, Vol. 10, 1975, pp. 234–242.

23 Odem, E. M. and Adams, D. F. "Failure Modes of Unidirectional Carbon/Epoxy Composite Compression Specimens," *Composites*, Vol. 21, No. 4, 1990, pp. 289–296.

24 Ashby, M. F. and Hallam, S. D. "The Failure of Brittle Solids Containing Small Cracks under Compressive Stress States," *Acta Metall.*, Vol. 34, No. 3, 1986, pp. 497–510.

25 Sammis, C. G. and Ashby, M. F. "The Failure of Brittle Porous Solids under Compressive Stress States," *Acta Metall.*, Vol. 34, No. 3, 1986, pp. 511–526.

26 Nemat-Nasser, S. and Horii, H. "Compression-Induced Nonplanar Crack Extension with Application to Splitting, Exfoliation, and Rockburst," *J. Geophys. Res.*, Vol. 87, No. B8, 1982, pp. 6805–6821.

27 Horii, H. and Nemat-Nasser, S. "Brittle Failure in Compression: Splitting, Faulting and Brittle-Ductile Transition," *Phil. Trans. R. Soc. Lond.*, Vol. A 319, 1986, pp. 337–374.

28 He, M.-Y. and Hutchinson, J. W. "Kinking of a Crack out of an Interface," *J. Appl. Mech.*, Vol. 56, 1989, pp. 270–278.

29 Oplinger, D. W., Parker, B. S., and Gandhi, K. R. *Studies of Tension Test Specimens for Composite Materials.* AMMRC Technical Publication #TR-82-87, 1980.

30 Oplinger, D. W., Parker, B. S., Foley, G., Lamothe, R., and Gandhi, K. R. "Comparisons of Tension Test Specimen Designs for Static and Fatigue Testing of Composite Materials," *International ASTM/Japan Meeting on Composites*, NASA-Langley, June 1983.

31 Starrett, S. "A Test Method for Tensile Testing Coated Carbon-Carbon and Ceramic Matrix Composites at Elevated Temperature in Air," *Ceram. Eng. Sci. Proc.*, Vol. 11, No. 9–10, 1990, pp. 1281–1294.

CHAPTER 6

Experimental Characterization and Theoretical Modeling of an Advanced Silicon Nitride Ceramic

J. L. DING,[1]

K. C. LIU[2] and C. R. BRINKMAN[2]

OVERVIEW

IN this chapter, some recent experimental results on creep and creep rupture behavior of a commercial grade of Si_3N_4 ceramic in the temperature range of 1150°C to 1300°C are summarized. In addition to comparing the data with some classical creep and creep rupture models, a tentative deformation and life prediction model for ceramic materials under general thermomechanical loadings is also proposed and evaluated. Some possible future research work to refine the design methodologies for high-temperature ceramics is also discussed.

INTRODUCTION

Recent improvements in materials processing technology have led to the use of advanced silicon nitride ceramics as feasible structural materials for many high-temperature engineering applications. Along with material development, a parallel development of new design methodologies is also necessary in order to mature the developed materials from research stage to practical use. For this purpose, an integrated data base and a realistic deformation and life prediction model for advanced ceramics at high temperatures are essential.

[1]Washington State University, Department of Mechanical and Materials Engineering, Pullman, WA, U.S.A.
[2]Oak Ridge National Laboratory, Metals and Ceramics Division, Oak Ridge, TN, U.S.A.

In the past, creep properties of ceramics were studied predominantly using flexure (bending) tests. Because of the nonuniform stress distribution in a specimen subjected to bending, extraction of accurate materials information from bending tests is quite difficult. For this reason, experimental data on the creep behavior of ceramics under pure uniaxial loading conditions is in critical need. As far as the model development is concerned, nearly all the reported data have been analyzed with some classical creep and creep rupture models such as use of Norton's power law relation [1] for steady-state creep to model creep behavior and the Larson-Miller model [2], minimum commitment method [3], or the Monkman-Grant relation [4] to model creep rupture behavior. While these relations may be useful for distinguishing creep and creep rupture behavior of different materials under ideal laboratory test conditions, i.e., isothermal and constant load test condition, they only have a limited capability for deformation and life prediction under general thermal mechanical loadings. In fact, Norton's power law relation may not even be able to describe the constant-stress creep behavior of most ceramics for which primary creep accounts for the major portion of the total creep.

In a recent study by Ding et al. [5], both creep and creep rupture behavior of a particular grade of Si_3N_4 were investigated systematically in the temperature range 1150°C to 1300°C in uniaxial tensile mode. Based on the obtained experimental data, a tentative phenomenological deformation and life prediction model for ceramics subjected to general thermal-mechanical loadings was also proposed [6]. This chapter is essentially an overall review of the key results obtained in References [5] and [6]. The emphases are on the characterization and modeling of the macroscopic behavior of structural ceramics at high temperatures, rather than the investigation of the detailed microscopic deformation and fracture mechanisms. Some possible future research work to refine the design methodologies for high-temperature ceramics is also discussed.

MATERIAL, SPECIMEN, AND EXPERIMENTAL APPARATUS

The material used in this study was a commercial grade of hot isostatically pressed (HIP) Si_3N_4, marketed as GN-10, which contains Y_2O_3 and SrO as densification aids.

Buttonhead tensile specimens having a 6.3-mm diameter and a uniform gage length of 25.4 mm were used for all creep experiments. Creep tests were performed on four standard lever-arm creep testing machines with the load applied through the dead weights. To maintain a uniform tensile stress in the test specimen, self-aligning gripping fixtures [7] were incorporated in the load-train assembly. A low-profile, two-zone–controlled resistance heating furnace capable of achieving temperatures to 1600°C was used to heat the specimen. The low heating profile is designed to heat only the center portion of the specimen so that "cold-gripping" at the specimen ends can be utilized. The hot zone inside the furnace is divided into top and bottom heating zones. Each zone is heated by a set of six $MoSi_2$ heating elements and controlled independently by a temperature controller. The temperature along the gage section was measured with three thermocouples, one placed near the middle and one near each end of the gage length; the thermocouple beads were not in direct contact with the test specimen. The temperature gradient between the center and the end of the gage section was less than 0.5 percent of the maximum temperature at the center. Creep strain was measured by a mechanical extensometer developed by Liu and Ding [8]. The extensometer has a resolution of 5 microstrain and an absolute accuracy better than 100 microstrain for long-term creep testing.

EXPERIMENTAL RESULTS

CREEP AND CREEP RUPTURE BEHAVIOR OF THE AS-HIPED MATERIAL

The matrix for the creep tests performed and resultant rupture times are shown in Table 6.1, and creep curves are shown in Figures 6.1 to 6.4. Sym-

TABLE 6.1. Matrix for Creep Tests of GN-10 Si_3N_4. The Number in Parentheses Indicates Duration of the Test, Which Is Denoted Either as Completed (X) or as Disrupted (D).

Stress	1150°C	1200°C	1250°C	1300°C
75 MPa			D^c(>2238 h)	D^d(>1125 h)
100 MPa			D^c(>1030 h)	X(1721 h)
125 MPa		D^a(>1031 h)	X(2996 h)	X(15.2 h)
150 MPa		X(1204 h)	X(135.9 h)	X^e(0.2 h)
175 MPa		D^b(>3405 h)	X(25.5 h)	
200 MPa		X(203.1 h)		
225 MPa		X(96.3 h)		
250 MPa	X(733.8 h)	X(7.5 h)		
275 MPa	No test			
300 MPa	X(365.4 h)			

[a]Stress increased to 225 MPa after 1031 h of testing.
[b]Fractured at specimen buttonhead due to a power outage.
[c]Fractured at specimen shank.
[d]Stress increased to 100 MPa after 1125 h of testing.
[e]No meaningful creep rate available due to fast fracture.

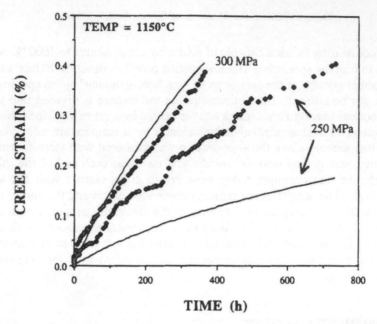

Figure 6.1 Creep curves of GN-10 Si₃N₄ tested at 1150°C. (Symbols are experimental data, and solid lines are predictions of the proposed model. The legend is used throughout this chapter unless specified otherwise).

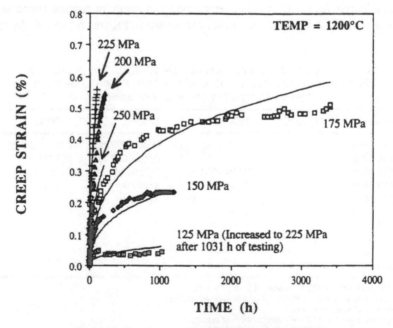

Figure 6.2 Creep curves of GN-10 Si₃N₄ tested at 1200°C.

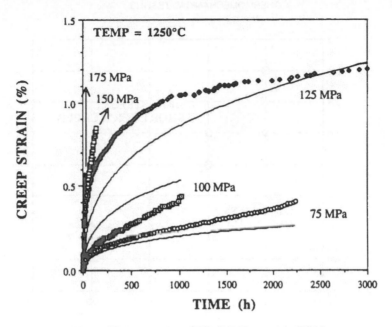

Figure 6.3 Creep curves of GN-10 Si₃N₄ tested at 1250°C.

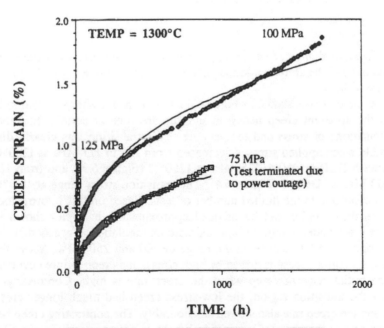

Figure 6.4 Creep curves of GN-10 Si₃N₄ tested at 1300°C.

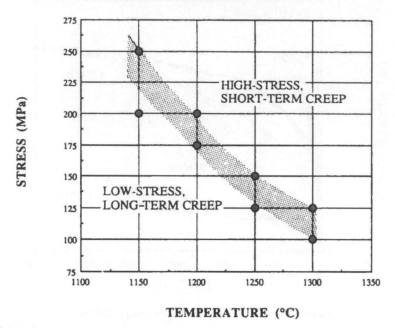

TEMPERATURE (°C)

Figure 6.5 The approximate transition region (shaded area), which separates the high-stress, short-term creep and low-stress, long-term creep.

bols in these figures represent experimental data and the solid lines are predictions of the aforementioned phenomenological model, which will be discussed later.

The creep curves shown in Figures 6.1 to 6.4 show a sharp transition in both the apparent creep behavior and rupture time at certain threshold combinations of stress and temperature. This transition was clearly discernible when applied stresses decreased from 200 to 175 MPa at 1200°C (Figure 6.2), from 150 to 125 MPa at 1250°C (Figure 6.3), and from 125 to 100 MPa at 1300°C (Figure 6.4). The transition stress range at 1150°C is not clear due to the limited number of tests. Based on this information, the transition region can be mapped approximately in a plot shown in Figure 6.5 (shaded area). An approximate extrapolation suggests that the transition at 1150°C will be in the range of 200 and 250 MPa. Above the transition region, creep is driven by high stress, and creep failure occurred in the initial stage of creep when the creep rate is high. Contrastingly, below the transition region, the low-stress creep had much longer creep life, and the creep rate also slowed considerably. The contrasting creep behavior causes the order-of-magnitude breaks in creep rate and creep lifetime within the transition region.

Another obvious feature of the overall creep behavior is the lack of tertiary creep. Tertiary creep is usually attributed to distributed damage caused by the nucleation and growth of cavities or microcracks, which reduces the load-bearing capacity of the specimen. The lack of tertiary creep may suggest that fracture was dominated by localized damage due to the growth of preexisting defects, e.g., macrocracks or voids, even though other damaging mechanisms may operate concurrently. In fact, very few cavities were observed in the as-HIPed specimen following creep [5].

Analysis of the creep rupture behavior indicates that creep rupture time, t_r, generally decreases as applied stress increases. Anomalies are noticed in three cases, where t_r values for both tests at 1250°C with 75 MPa and 100 MPa were lower compred to that tested at 125 MPa; similarly, t_r at 1200°C and 150 MPa was lower compared to that at 175 MPa. The first two specimens fractured in the shank, as noted in Table 6.1. In the latter case, the premature fracture is believed to be due to an inhomogeneous distribution of defects. In general, some scatter in t_r as a function of stress is not unusual, especially for ceramic materials, for which fracture may be sensitive to the distribution and characteristics of the defects.

EFFECTS OF ANNEALING (AGING) ON SUBSEQUENT CREEP BEHAVIOR

In ceramics, low dislocation activity occurring inside the matrix grains does not contribute significantly to creep deformation [9,10]. The major strengthening mechanism in GN-10 Si_3N_4 was probably due to hardening of the grain boundary phase by high-temperature annealing. Additional details describing likely strengthening mechanisms operative in Si_3N_4 during primary or transient creep can be found elsewhere [11]. Figure 6.6 compares the creep behavior of the as-HIPed material tested at 1200°C and 225 MPa and that of a specimen precrept at 125 MPa for 1031 h prior to the application of additional load to 225 MPa. To facilitate comparison, the zero time for the precrept specimen was reset at the beginning of the second loading to 225 MPa. Although the precrept specimen fractured prematurely at the buttonhead due to a power outage, the enhancement in creep resistance exhibited by the precrept specimen was clearly discernible. Since the preceding low-stress creep produced little or no creep strain after the initial transient creep (see Figure 6.2), the subsequent strengthening must be credited to prolonged thermal exposure at high temperature. Similar findings were reported by Cranmer et al. [12] and Todd and Xu [13]. The hardening of grain boundary phase occurring during long-term creep could be attributed to increasing nitrogen concentration levels in the oxynitride glass [13].

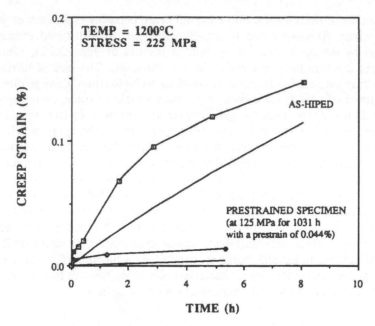

Figure 6.6 Comparison of the initial transient creep behavior of the as-HIPed specimen tested at 1200°C with 225 MPa and that of a precept specimen subjected to the same test condition but preceded by initial testing at 1200°C and 125 MPa for 1031 h.

To further study the "hardening" effect due to thermal annealing, four specimens were annealed at 1370°C in air for 150 h. After annealing, a thin layer of oxide scale was found on the specimen surface [14]. The oxide scale was removed, and the specimen surface was polished with wet SiC paper before testing. Creep curves of the annealed and the as-HIPed specimens were compared for each temperature tested at 1150, 1200, 1250, and 1300°C. A comparison of the creep data obtained at 1200°C is shown in Figure 6.7 to illustrate the contrast in creep behavior. Quantitative comparisons were also made between the annealed and as-HIPed properties in three categories of total creep strain at failure, creep rate at fracture, and rupture time in terms of percentage, and results are tabulated in Table 6.2. Thermal annealing was clearly effective for the enhancement of both creep resistance and rupture life of GN-10 Si_3N_4. The low percentage of rupture-life for the 1150°C/300 MPa test due to premature failure of the annealed specimen misrepresents the general trend. Relevant studies on the effects of crystallization of grain boundary glass phase on the strengthening of Si_3N_4 can also be found in References [14] and [15].

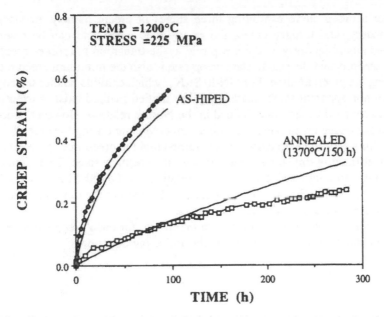

Figure 6.7 Comparison of the creep behavior of annealed and the as-HIPed specimens tested at 1200°C and 225 MPa.

CLASSICAL ANALYSIS OF CREEP AND CREEP RUPTURE DATA

ANALYSIS OF CREEP DATA BY NORTON'S POWER LAW RELATION

The Norton power law relation is described by

$$\dot{\epsilon} = A\sigma^n e^{-(Q/RT)} \tag{1}$$

where $\dot{\epsilon}$ is the steady-state creep rate in h^{-1}, σ the stress in MPa, T the absolute temperature in K, R the gas constant, Q the activation energy in kJ/mole, n the stress exponent, and A a material constant.

TABLE 6.2. Percentages of Creep Strain, Creep Rate at Fracture, and Rupture Time of the Annealed Specimens Relative to Those of the as-HIPed Specimens Tested at the Conditions Indicated.

	1150°C 300 MPa	1200°C 225 MPa	1250°C 175 MPa	1300°C 125 MPa
Creep strain	19%	43%	74%	69%
Creep rate at fracture	66%	15%	12%	14%
Rupture time	18%	295%	444%	430%

For a creep curve exhibiting three stages of creep, namely, primary, secondary, and tertiary creep, the steady-state creep rate can be determined unambiguously as the creep rate at the transition to tertiary creep. It is obvious that the steady-state creep rate is also the minimum creep rate during the creep lifetime. For GN-10 Si_3N_4, which exhibits neither tertiary creep nor apparent steady-state creep, no unified method exists currently to determine the creep rate defined in the Norton relation. However, since the steady-state creep rate is the minimum creep rate during the creep lifetime for a typical creep curve, it is assumed that the creep rate is the minimum at fracture time for a creep curve with no tertiary stage. This method was used to characterize the Norton relation for GN-10 Si_3N_4. Accordingly, the term in Equation (1) should be understood as the minimum creep rate instead of the steady-state creep rate.

In order to calculate the minimum creep rate for each creep curve, experimental data were first fitted by the following equation:

$$\epsilon = x_1[(1 + x_2t)^{x_3} - 1] \qquad (2)$$

where ϵ is the creep strain; t the time in h; and x_1, x_2, and x_3 coefficients to be determined by curve fitting. The minimum creep rate for each curve was then calculated from the time derivative of Equation (2) at fracture time. Calculated minimum creep rates are plotted in Figure 6.8, in which open and filled-in symbols were used to represent the data above and below the aforementioned transition region, respectively. For the incomplete tests, including those in which specimens fractured in the shank, a downward arrow was attached to the data point to imply that the actual creep rate might be lower than indicated. Note that the specimen tested at 1300°C/150 MPa failed in a short time, like a tensile test. In this case, a meaningful creep rate cannot be determined, as noted in Table 6.1.

The data points are clearly polarized in two groups according to symbols and separated by an area devoid of data, which can be readily identified with the transition region discussed in Figure 6.6. Creep data for the four annealed specimens are not included in Figure 6.8 because their creep behavior had been modified by postexposure to elevated temperature. A separate analysis would be more appropriate for the annealed material, but impractical in this case due to the limited information.

A multivariate regression analysis performed on all the creep data, except for those with an arrow attached, yields the following expression for the Norton's power law relation:

$$\dot{\epsilon} = e^{56}\sigma^{12.6}e^{-(1645/RT)} \qquad (3)$$

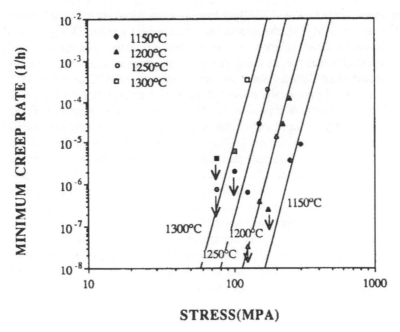

STRESS(MPA)

Figure 6.8 Comparison of the minimum creep rate data and the Norton power law equation (straight lines), with stress exponent and activation energy equal to 12.6 and 1645 kJ/mole, respectively. Open and filled-in symbols represent the data above and below the experimentally observed transition region, respectively. Arrowed data implies the actual creep rate may be lower than the indicated one.

where the values of n and Q in Equation (1) are estimated to be 12.6 and 1645 kJ/mole, respectively. A comparison between the creep rate data and Equation (3) represented by solid lines is shown in Figure 6.8. In deriving Equation (3), three filled-in data (without arrow) were used. However, since these three data were on the boarder of the transition region, 12.6 and 1645 kJ/mole for n and Q, respectively, as obtained above, essentially represent the creep behavior in high-stress regime. Some other possible values for n and Q based on different approaches for selecting the data were discussed in Reference [16] whose scope is very similar to this chapter.

ANALYSIS OF CREEP RUPTURE DATA BY CLASSICAL CREEP RUPTURE MODELS

Examples of classical creep rupture models for modeling creep rupture data under isothermal and constant loading conditions include the Larson-Miller model (LM) [2], the minimum commitment method (MC) [3], and the Monkman-Grant relation [4], as mentioned earlier. All three models were utilized to analyze the creep rupture data given in Table 6.1, including the rupture time data for the specimen tested at 1300°C and 150 MPa, which, as noted previously, provided no useful information in Figure 6.8 to aid in the analysis of Equation (1). The arrowhead data shown in Figure 6.9 are plotted likewise in Figure 6.8, implying that the rupture time would be longer if premature failure had not occurred at the time indicated. Examination of Figure 6.2 reveals that the specimen tested at 1200°C and 150 MPa followed the characteristic trend of creep behavior, i.e., low stress producing low creep rate, but its short creep life compared to that for the specimen tested at 175 MPa was rather contrary to the general trend, which usually showed that high applied stress gave short life and vice versa. Therefore, this data point was considered to be an anomaly and tagged with a solid arrowhead, as shown in Figures 6.9 and 6.10, to indicate it was censored also. The arrowed data were again excluded in the following analyses but are plotted in Figures 6.9 and 6.10 for comparisons with pre-

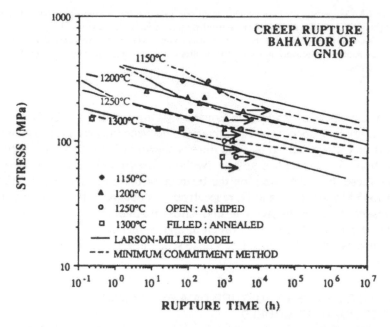

Figure 6.9 Comparison of experimental data and predictions of the Larson-Miller model and minimum commitment method. Arrowed data were not included in the regression analysis.

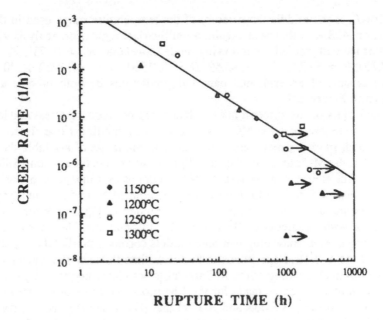

RUPTURE TIME (h)

Figure 6.10 Comparison of experimental data and the Monkman-Grant relation represented by a straight line. Arrowed data were not included in the regression analysis.

dictions to examine if the data points fall in logical locations relative to the predictions. Except for the aforementioned two data points, the arrowed data set for creep rupture analysis is essentially the same as that for deriving Equation (3).

The Larson-Miller model is described by the following equation:

$$\log t_r = B_0 + \frac{B_1}{T} + \frac{B_2}{T} \log \sigma \tag{4}$$

where B_0, B_1, and B_2 are constants. A multivariate regression analysis was performed on the non-arrowed data in Figure 6.9 and yielded the model constants as $B_0 = -58.28$, $B_1 = 136,600$ and $B_2 = -20,510$.

The minimum commitment method gives an equation of the following form:

$$\log t_r + \left[R_1(T - T_m) + R_2 \left(\frac{1}{T} - \frac{1}{T_m} \right) \right] = B + C \log \sigma + D\sigma + E\sigma^2 \tag{5}$$

where T_m is the middle temperature of the temperature range used in the tests, or 1498 K in this case. Again, a multivariate regression analysis was performed and yielded the following model constants: $R_1 = 0.1731$, $R_2 = 303{,}700$, $B = 87.24$, $C = -45.69$, $D = 0.11560$, $E = -0.9733 \times 10^{-4}$. Comparisons of experimental data and predictions by both models are shown in Figure 6.9.

The LM model describes a linear relationship between log stress and log rupture time, whereas the MC model describes a nonlinear one. Interestingly, both models predict about the same in the range of available data. However, the MC model fits the data slightly better than the Larson-Miller model. This is expected because the MC model has six independent variables, as opposed to three for the LM model. Four filled-in data points representing the annealed specimens are also plotted in Figure 6.9 for comparisons with predictions. Generally, the filled-in data fall above the predictive curves, indicating that both models correctly predict the rupture lives of the as-HIPed specimens to be lower than those of the annealed.

The physical meaning of the MC model is not clear; however, a plausible interpretation can be made for the LM model in terms of macrocrack growth. According to Evans and Blumenthal [17], when the crack growth rate, $\dot{a}$, of a semicircular surface crack or a penny-shaped internal crack with radius a is characterized by

$$\dot{a} = \dot{a}_0 \left(\frac{K}{K_c}\right)^m e^{-(Q/RT)} \tag{6}$$

where $\dot{a}_0$ is a constant, K the stress intensity factor, K_c the critical stress intensity factor, and m an exponent comparable to that of n in Equation (1), the time for the crack to grow from its initial size to the critical size, i.e., the rupture time, is proportional to σ^{-m}. The same relationship was derived by Lange [18]. When Equation (4) is written in a power function of rupture time, m in Equation (6) is equal to $-B_2/T$ in Equation (4). Using $B_2 = -20{,}510$ derived from the regression analysis, one finds that m varies from 14.4 at 1150°C to 13 at 1300°C. Both values are comparable to the stress exponent $n = 12.6$ given in Equation (3).

The Monkman-Grant model describes a power law relation between the rupture time and the minimum creep rate, i.e.,

$$t_r = A\dot{\epsilon}^{-p} \tag{7}$$

The constants A and p were determined to be 0.0021 and 0.91, respectively, using the nonarrowed data points shown in Figure 6.10. The data points having an arrow were not used due to the incomplete information. Data and the solid line representing Equation (7) are compared in Figure 6.10.

The Monkman-Grant relation is analogous to the Larson-Miller model because substitution of Equation (1) into Equation (6) leads to a power law relation for rupture time in terms of stress. A comparison of the derived power law relation with the LM equation will reveal that the stress exponents m, n, and p are interrelated as $p = m/n$. According to the values of m and n determined in preceding sections, p should fall in the range of 1.03 to 1.14, which is slightly higher than the value of p representing the slope of the Monkman-Grant curve illustrated in Figure 6.9. Independently, Wiederhorn et al. [19] and Ohji et al. [20] reported values of p to be 0.96 and 1, respectively, for their respective Si_3N_4 ceramics. The agreement among the three p values is respectable.

A TENTATIVE DEFORMATION AND LIFE PREDICTION MODEL FOR GENERAL THERMAL-MECHANICAL LOADINGS

DEVELOPMENT OF THE MODEL

As mentioned earlier, although the classical creep and creep rupture models may give reasonable description of the experimental data under ideal laboratory test conditions, i.e., isothermal and constant-load conditions, they, in general, have very limited capabilities for practical applications in which both temperature and load vary with time. Therefore, in order to effectively evaluate the mechanical reliability of ceramic components, a more realistic deformation and life prediction model for general thermal-mechanical loadings is necessary.

For ceramics, there have been some research efforts devoted to the development of advanced models. For deformation prediction, some examples can be found in the work by Lange [21], Dryden et al. [22], and Chadwick et al. [23]; for fracture prediction, examples are the work by Evans and Blumenthal [17], Tsai and Raj [24], and Chuang [25]. The purpose of these so-called micromechanical models is, in general, to understand better the contribution to deformation and fracture of ceramics due to certain specific deformation and fracture mechanisms. These specific models are usually quite complicated themselves. It is almost a formidable task to attempt to further incorporate all the possible deformation and fracture mechanisms to predict the overall or averaged macroscopic material behavior. For engineering design purpose, a simplified phenomenological continuum model is therefore more appropriate. Based on the aforementioned behavior and microstructural features of this particular grade of ceramic, a tentative phenomenological model was proposed by visualizing the material as an aggregate of Si_3N_4 grains, which are bonded by a grain boundary phase material with some preexisting defects in the forms of surface microcracks and internal voids in the bulk. More specifically, the proposed phenomenological model was formulated on the basis of the following assumptions:

(1) Creep rate is a nonlinear function of stress, temperature, and the current inelastic state of the material. The time dependence may be mainly attributed to the viscosity of the residual amorphous phase in the grain boundaries.

(2) Hardening, which increases the resistance to both creep deformation and damage propagation, is assumed to be mainly due to the progressive devitrification of the residual amorphous phase, and enhanced as temperature and soaking time increase.

(3) Because few cavities were observed in the as-HIPed specimens, creep fracture may be mainly dominated by the growth of macrocracks or voids. The lack of accelerated or tertiary creep may also suggest that material fractures before the damaging mechanisms, due to either crack propagation or cavity nucleation and growth, have significant effects on creep deformation. Therefore, for the current material, it is assumed that there is little interaction between creep deformation and creep damage.

Formulation of the model was based on the state variable approach whose general background can be found in the work by Rice [26] and others. Two internal state variables, namely, a hardening variable (δ) and a damage variable (ω), were employed to characterize the current state of the material. The model consists of three equations—a flow rule that describes the strain rate ($\dot{\epsilon}$) as a function of the state variables, applied stress (σ), and temperature—and two evolution rules for the two internal state variables.

Specifically, a model of the following form is proposed:

$$\dot{\epsilon} = \frac{d\epsilon}{dt} = \frac{f(\sigma,T)\dot{\epsilon}_0 e^{-(Q_\epsilon/RT)}}{\delta} = \frac{\alpha}{\delta} \tag{8}$$

$$\dot{\delta} = \frac{d\delta}{dt} = \frac{\dot{\delta}_0}{\delta^m} e^{-(Q_\delta/RT)} = \frac{\beta}{\delta^m} \tag{9}$$

$$\dot{\omega} = \frac{d\omega}{dt} = \frac{\dot{\omega}_\sigma \left(\frac{\sigma}{\sigma_0}\right)^\nu \epsilon^{-(Q_\omega/RT)}}{\delta(1-\omega)} = \frac{\gamma}{\delta(1-\omega)} \tag{10}$$

where

$$\alpha = f(\sigma,T)\dot{\epsilon}_0 e^{-(Q_\epsilon/RT)} \tag{11}$$

$$\beta = \delta_0 e^{-(Q_\delta/RT)} \tag{12}$$

$$\gamma = \dot{\omega}_0 \left(\frac{\sigma}{\sigma_0}\right)^\nu e^{-(Q_\omega/RT)} \tag{13}$$

f is a function of σ and T, Q_ϵ, Q_δ, Q_ω, $\dot{\epsilon}_0$, $\dot{\delta}_0$, m, $\dot{\omega}_0$, and ν are constants. Time t is in hour, strain ϵ in microstrain, and stress σ in MPa.

The initial value of the hardening variable (δ) is assumed to be 1 and increases as the level of devitrification increases subsequently. Equation (1) therefore implies that the initial creep rate of the as-HIPed specimen at the inception of creep loading is equal to α. The creep rate decreases as δ increases. The damage variable is assumed to vary from 0 at the initial state of the material to 1, which corresponds to $\dot{\omega} = \infty$ at fracture. The temperature dependence of all the rate equations are assumed to be of the Arrhenius form, i.e., $e^{-(Q/RT)}$.

Equation (9) indicates that the rate of devitrification is a function of soaking temperature and slows inversely as the level of devitrification increases. Under assumption 2, Equation (9) is coupled with Equations (8) and (10), which, however, are independent of each other because of assumption 3. With no introduction of δ being associated with material hardening, due to devitrification in this case, and decoupling of ϵ and ω, Equations (8) and (10) are often used to describe the transition behavior from secondary to tertiary creep of metals [27].

CHARACTERIZATION OF THE MODEL

The function f and the material constants in Equations (8) and (9) can be determined by the use of creep data and those in Equation (10) by the creep rupture life data from both as-HIPed and annealed specimens. The details of the model characterization are described in Reference [6]. The function f was determined to be of the following form:

$$f(\sigma, T) = (\sigma_0)^n \left(\frac{\sigma - \sigma_{th}}{\sigma_0} - c\right)^n \tag{14}$$

where σ_{th} is the threshold stress, below which the creep is assumed to be negligible, given by

$$\sigma_{th} = 1765.8 - 1.12T \quad \text{for } T \leq 1250°C \tag{14a}$$

and

$$\sigma_{th} = 516.9 - 0.3T \quad \text{for } T > 1250°C \tag{14b}$$

Material constants c and n take different values in different sectors of the stress-temperature space, shown in Figure 6.5. Results of a detailed analysis performed in Reference [6] showed that transition zone (shaded area) reduced to the following bilinear expressions

$$\sigma_{trans} = 1848.3 - 1.12T \quad \text{for } T \leq 1250°C \quad (15a)$$

and

$$\sigma_{trans} = 1108.1 - 0.634T \quad \text{for } T > 1250°C \quad (15b)$$

Equation (15a) falls approximately on the upper bound, and Equation (15b) indicates σ_{trans} at 1300°C falls in the middle of the band.

For $\sigma \leq \sigma_{trans}$, $c = 0$ and $n = 1$

$\sigma > \sigma_{trans}$, $c = 1$ and $n = 1.32$ for $T \leq 1250°C$

$c = 1$ and $n = 1.7$ for $t > 1250°C$

Remaining material constants are determined to be

$\sigma_0 = 54$

$m = 1/3$

$\dot{\epsilon}_0 = e^{78.06}$ for $T \leq 1200°C$ $e^{21.929}$ for $T > 1200°C$

$Q_\epsilon = 957.4$ kJ/mole for $T \leq 1200°C$ 270 kJ/mole for $T > 1200°C$

$\delta_0 = e^{93.5}$ for $T \leq 1200°C$ $e^{-1.26}$ for $T > 1200°C$

$Q_\delta = 1174$ kJ/mole for $T \leq 1200°C$ 13.26 kJ/mole for $T > 1200°C$

$\dot{\omega}_0 = e^{103.46}$, $v = 10.47$, and $Q_\omega = 1497$ kJ/mole

EVALUATION OF THE MODEL

Prediction of Creep Deformation

Isothermal, Constant Stress Creep

Under isothermal and constant stress conditions, a closed form solution can be obtained for the creep strain as a function of time, as described by the following equation:

$$\epsilon = \frac{\alpha}{m\beta} ((1 + (1 + m)\beta t)^{m/(1+m)} - 1) \quad (16)$$

Creep curves predicted by Equation (16) are shown as the solid lines in Figures 6.1 to 6.4. Examinations of these figures indicate that the essential features of creep behavior are reasonably well described in each case with consideration of the temperature and stress ranges that were covered. It should also be noted that Equation (16) has the same form as Equation (2).

Effects of Annealing on Subsequent Creep Behavior

For the annealed specimens, a closed form solution for creep strain can again be derived from the model as follows:

$$\epsilon = \frac{\alpha}{m\beta} (((1 + m)\beta t + \bar{\delta}^{m+1})^{m/(1+m)} - \bar{\delta}^m) \tag{17}$$

where $\bar{\delta}$ is the value of the devitrification variable at the end of annealing given by

$$\bar{\delta} = (1 + (1 + m)\beta_{1370°C} \cdot 150)^{1/(1+m)} \tag{18}$$

An example of predicted results (solid line) based on Equation (17) is shown in Figure 6.7, which demonstrates that annealing effect on the subsequent creep behavior is well demonstrated by the model.

Creep under Stepwise Varied Loading

Since the model assumes no strain hardening, the creep strain at each loading step can also be calculated using Equation (17), with $\bar{\delta}$ substituted by the value of δ evaluated at the end of the previous loading step. The total creep strain is the accumulation of the creep strain resulting from each loading step. Comparisons of calculated and experimental creep curves for a specimen subjected to a single stepped load (depicted in Figure 6.6) show that essential behavioral features are in qualitative agreement.

Loading at 1300°C is shown in Figure 6.11. As indicated, the first leg of testing at 75 MPa was interrupted at $t = 937$ h due to a power failure. Although the specimen cooled down to ambient temperature, the subsequent creep behavior did not appear to have been altered when testing resumed. The load was increased to 100 MPa after completing 1125 h of testing and further increased to 125 MPa at $t = 1437$ h, until specimen fracture occurred at $t = 1528.82$ h. The solid line indicates the predicted creep curve, which agrees well with the first segment of the experimental data, but progressively underestimates the remaining data as the applied stress steps upward. In view of the poor consistency in creep behavior generally reported in the open literature due to material anomalies and batch variation, the simulation exhibited in Figure 6.11 is respectable in the qualitative sense.

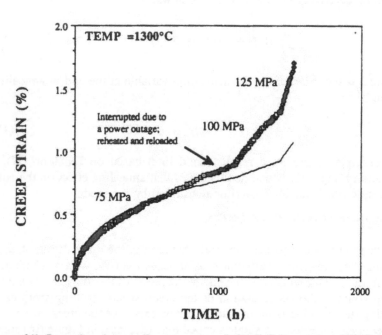

Figure 6.11 Comparison between the creep curve of GN-10 Si_3N_4 tested at 1300°C under stepwise-varied loading and the solid line predicted by the proposed model.

Prediction of Creep Rupture

Isothermal, Constant Stress Creep Rupture

Under isothermal and constant stress conditions, the model leads to the following equation for rupture time for both as-HIPed and annealed specimens:

$$t_r = \frac{1}{(1 + m)\beta} \left(\left(\frac{m\beta}{2\gamma} + k^m \right)^{(m+1)/m} - k^{m+1} \right) \qquad (19)$$

where k is the initial value of δ at the inception of creep loading. Figure 6.12 shows that predicted rupture-time curves, solid lines for the as-HIPed and dashed for the annealed specimens, compare quite well with experimental data. It is interesting to note that predicted rupture-time curves for each temperature in the pair merge together as the rupture times increase. This observation implies that the effects of annealing on rupture time diminish if tests are run at low stresses over long periods of time.

Creep Rupture under Stepwise Varied Loading

Under stepwise varied loading conditions, it can be shown that, for each loading step, ω can be calculated from the following expression:

$$-\frac{1}{2}(1 - \omega)^2 = \frac{\gamma}{\beta m}((1 + m)\beta t + k^{1+m})^{m/(1+m)} - \frac{1}{2}(1 - \bar{\omega})^2 - \frac{\gamma}{m\beta} k^m$$

$$(20)$$

where t is the time expended at a given step load and $\bar{\omega}$ is the initial value of ω when the step load was applied to the specimen. An attempt was made to estimate the rupture time of the specimen tested in the conditions delineated in Figure 6.11, using Equation (20). Values of $\omega = 0.035, 0.037,$ and 0.106 were obtained at $t = 937, 1125,$ and 1437 h, respectively. Under the last step of loading at 125 MPa, the model predicts a rupture time of 215 h, which overpredicts the actual rupture time of 92 h by 123 h. For life prediction of ceramic materials under variable loadings, an overestimation of 123 h out of actual rupture life of 1529 h is not excessive.

FUTURE RESEARCH WORK

Compared to the experimental data, although the model did not fit the data point by point, it did reasonably describe all the essential features of creep and creep rupture behavior under both constant and variable loadings, considering the range of the stress and temperature and types of load-

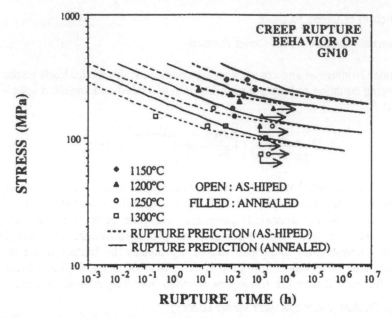

Figure 6.12 Comparison between experimental creep rupture times of both the as-HIPed and annealed specimens and theoretical predictions.

ing that experimental data covered. The model can also estimate the effects of annealing on creep and creep rupture. It is essentially a design tool that can, at least, give a first approximation of both creep deformation and rupture life of ceramics under general thermal mechanical loadings as encountered in practical applications. Its capability is beyond the combination of traditional Norton's creep law and Larson-Miller or Monkman-Grant rupture models.

As demonstrated above, the model is simple enough to render closed form solutions for creep deformation and rupture life at least for stepwise varied loadings. Furthermore, it has been shown that determination of the model constants or characterization of the model is also very straightforward. These features should allow design engineers to conveniently characterize a new material and carry out the preliminary design without the necessity of using a computer. Of course, with the same approach, a better fit of the experimental data could be obtained by adopting a more complicated mathematical form. However, in this case, the determination of the material constants has to use some iterative procedure, and the evaluation of the model has to be carried out with numerical integration by computer.

The more pressing issues for refining the design methodologies for ceramics seem to be the following.

DEVELOPMENT OF THE STOCHASTIC MULTIAXIAL DEFORMATION AND LIFE PREDICTION MODEL

Although the model described previously is capable of predicting the creep and creep rupture behavior under general thermal-mechanical loadings, its present scalar form limits its applicability only to uniaxial stress states such as simple tension. Since nearly all the engineering components are subjected to complicated stress states in practical applications, the model must be extended to the tensor form in order to be truly useful for mechanical reliability analysis. The tensor equation must also take into consideration the possible asymmetry of the creep and creep rupture behavior under tension and compression. For metals, a lot of research, in both theories and experiments, has been done to investigate the possible extension of scalar to tensor equations. To the authors' knowledge, a parallel work for ceramics is virtually nonexistent at this point.

As the distribution of fracture-initiating defects in ceramics are random, an ultimate model must also take into account the volume effect as well as the stochastic nature of the creep rupture behavior. For example, in a deterministic approach, as described in the proposed model, fracture is assumed to initiate at the defect under the highest principal stress. However, in reality, the occurrence of fracture depends on the distribution, size, and shape of defects and could be highly stochastic in nature. Therefore, provisions to account for the statistical diversity of creep and creep rupture behavior must be considered in the future model to make reliability analysis practical. To this end, a much wider data base is needed.

ADVANCED EXPERIMENTATION ON CREEP AND CREEP RUPTURE BEHAVIOR OF CERAMICS

With the increasing maturity of the uniaxial testing techniques, future research efforts in material testing should focus on the study of creep behavior of ceramic materials under multiaxial stress states. These data are essential for model validation and refinement work. Some biaxial studies of ceramic materials at room temperature have been reported [28]. But only limited multiaxial creep data are currently available for developing a multiaxial material model.

It should also be noted that constant stress and temperature are the ideal laboratory test conditions. Experimental data of this type are of fundamental importance to material characterization. However, stress and temperature usually vary with time in practical applications. Experimental data simulating the operating condition are also desirable in order to gain the insight of material behavior and to evaluate the theoretical model. Since not all field conditions can be reproduced in laboratory, data may be obtained under somewhat modified conditions. Nevertheless, information such as that obtained under stepwise-varied loading conditions can be extremely useful.

CONCLUSION

In this chapter, a comprehensive set of experimental data on creep and creep rupture behavior of a commercial grade of Si_3N_4 are presented. Lack of tertiary creep and the order of magnitude breaks in creep rate and rupture lifetime at certain threshold combinations of stress and temperature were the two characteristic features of the creep behavior of GN-10 Si_3N_4 at temperatures in the range 1150 to 1300°C. It was also found that thermal annealing (aging) enhanced both subsequent creep resistance and creep rupture life of GN-10 Si_3N_4 and might be a major hardening mechanism during creep.

In addition to evaluating several classical creep and creep rupture models, a deformation and life prediction model for ceramics under general thermal-mechanical loadings was also proposed and evaluated with reference to the obtained data. Although the proposed model is exploratory in nature, it has demonstrated the capability of describing the essential features of uniaxial creep and creep rupture behavior of the material under both constant and stepwise-varied loading conditions. Introduction of hardening and damage variables in the model has further enhanced its ability to predict the effects of annealing on creep behavior and creep rupture response. Simplicity of the model also makes the closed form solutions for constant or stepwise-varied loading possible.

Discussions were given concerning the needs of future theoretical and experimental research work for model refinement to include features such as multiaxiality, asymmetry of creep in tension and compression, and stochastic nature of defects inherent to ceramic materials.

ACKNOWLEDGEMENTS

This research was sponsored by the U.S. Department of Energy (USDOE), Assistant Secretary for Conservation and Renewable Energy, Office of Transportation Technologies, as part of Ceramic Technology Project of Materials Development Program, under contract DE-AC05-84OR21400 with Martin Marietta Energy Systems, Inc.

J. L. Ding would also like to acknowledge the partial support provided by the Faculty Research Participation Program administered by Oak Ridge Associated Universities (ORAU).

REFERENCES

1 Norton, F. H. *The Creep of Steel at High Temperatures*. McGraw-Hill, 1929.

2 Larson, F. R. and Miller, J. "A Time-Temperature Relationship for Rupture and Creep Stresses," *Trans. of ASME*, Vol. 74, 1952, pp. 765–771.

3 Manson, S. S. and Muralidharan, U. "Analysis of Creep Rupture Data for Five Multiheat Alloys by the Minimum Commitment Method Using Double Heat Term Centering Technique," *Progress in Analysis of Fatigue and Stress Rupture*, ASME, 1984, pp. 1–46.

4 Monkman, F. C. and Grant, N. J. "An Empirical Relationship between Rupture Life and Minimum Creep Rate in Creep-Rupture Test," *Proc. ASTM*, Vol. 56, 1956, pp. 593–620.

5 Ding, J. L., Liu, K. C. and Brinkman, C. R. "Creep and Creep Rupture of an Advanced Silicon Nitride Ceramic," *J. Am. Ceram. Soc.*, Vol. 77, No. 4, 1994, pp. 867–874.

6 Ding, J. L., Liu, K. C. and Brinkman, C. R. "Development of a High Temperature Deformation and Life Prediction Model for an Advanced Silicon Nitride Ceramic," *J. Am. Ceram. Soc.*, in press.

7 Liu, K. C. and Brinkman, C. R. "Tensile Cyclic Fatigue of Structural Ceramics," *Proc. 23rd Automotive Technology Development Contractors' Coordination Meeting*, October 21–24, 1985, Dearborn, Michigan, P-165, SAE, Warrendale, PA, 1986, pp. 279–283.

8 Liu, K. C. and Ding, J. L. "A Mechanical Extensometer for High Temperature Tensile Testing of Ceramics," *J. of Testing and Evaluation*, Vol. 21, No. 5, 1993, pp. 406–413.

9 Evans, A. G. and Sharp, J. V. "Microstructural Studies on Silicon Nitride," *J. Mater. Sci.*, Vol. 6, 1971, pp. 1292–1302.

10 Kossowsky, R. "The Microstructure of Hot-Pressed Silicon Nitride," *J. Mater. Sci.*, Vol. 8, 1973, pp. 1603–1615.

11 Wiederhorn, S. M., Hockey, B. J. and Cranmer, D. C. "Transient Creep Behavior of Hot Isostatically Pressed Silicon Nitride," *J. Mater. Sci.*, Vol. 28, 1993, pp. 445–453.

12 Cranmer, D. C., Hockey, B. J., and Wiederhorn, S. M. "Creep and Creep-Rupture of HIP-ed Si_3N_4," *Ceram. Eng. Sci. Proc.*, Vol. 12, No. 9–10, 1991, pp. 1862–1872.

13 Todd, J. A. and Xu, Z.-Y. "The High Temperature Creep Deformation of $Si_3N_4-6Y_2O_3-2Al_2O_3$," *J. Mater. Sci.*, Vol. 24, 1989, pp. 4443–4452.

14 Tsuge, A., Nishida, K., and Komatsu, M. "Effect of Crystallizing the Grain Boundary Glass Phase on the High Temperature Strength of Hot-Pressed Si_3N_4 Containing Y_2O_3," *J. Am. Ceram. Soc.*, Vol. 58, No. 7–8, 1975, pp. 323–326.

15 Cinibulk, M. K. and Thomas, G. "Grain Boundary Phase Crystallization and Strength of Silicon Nitride Sintered with a YSiAlON Glass," *J. Am. Ceram. Soc.*, Vol. 73, No. 6, 1990, pp. 1606–1612.

16 Ding, J. L., Liu, K. C., and Brinkman, C. R. "A Comparative Study of Existing and Newly Proposed Models for Creep Deformation and Life Prediction of Si_3N_4," *Life Prediction Methodologies and Data for Ceramic Materials*, C. R. Brinkman and S. F. Duffy, eds., *ASTM STP 1201*, ASTM, Philadelphia, PA, 1994, pp. 62–83.

17 Evans, A. G. and Blumenthal, W. "High Temperature Failure Mechanisms in Ceramic Polycrystals," *Deformation of Ceramics II*. R. E. Tressler and R. C. Bradt, eds., Plenum Press, New York, 1984, pp. 487–505.

18 Lange, F. F. "Interrelations between Creep and Slow Crack Growth for Tensile Loading Conditions," *Int. J. Fracture*, Vol. 12, No. 5, 1976, pp. 739–744.

19 Wiederhorn, S. M., Krause, R., and Cranmer, D. C. "Tensile Creep Testing of Structural Ceramics," *Proc. Annual Automotive Technology Development Contractors' Coordination Meeting*, October 23–31, 1991, Dearborn, Michigan, P-256, SAE, Warrendale, PA, 1992, pp. 273–280.

20 Ohji, T. and Yamauchi, Y. "Tensile Creep and Creep Rupture Behavior of Monolithic and SiC Whisker Reinforced Silicon Nitride Ceramics," *J. Am. Ceram. Soc.*, Vol. 72, No. 12, 1993, pp. 3105–3112.

21 Lange, F. F. "Non-elastic Deformation of Polycrystals with a Liquid Boundary Phase," *Deformation of Ceramic Materials*, R. C. Bradt and R. E. Tressler, eds., 1975, pp. 361–381.

22 Dryden, J. R., Kucerovsky, D., Wilkinson, D. S., and Watt, D. F. "Creep Deformation Due to a Viscous Grain Boundary Phase," *Acta Metall.*, Vol. 37, 1989, pp. 2007–2015.

23 Chadwick, M. M., Wilkinson, D. S., and Dryden, J. R. "Creep Due to a Non-Newtonian Grain Boundary Phase," *J. Am. Ceram. Soc.*, Vol. 75, No. 9, 1992, pp. 2327–2334.

24 Tsai, R. L. and Raj, R. "Creep Fracture in Ceramics Containing Small Amounts of Liquid Phase," *Acta Metall.*, Vol. 30, 1982, pp. 1043–1058.

25 Chuang, T. J. "A Diffusive Crack-Growth Model for Creep Fracture," *J. Am. Ceram. Soc.*, Vol. 65, No. 2, 1982, pp. 93–103.

26 Rice, J. R. "Inelastic Constitutive Relations for Solids: An Internal Variable Theory and Its Application to Metal Plasticity," *J. Mech. Phys. Solids*, Vol. 19, 1971, pp. 433–455.

27 Rides, M., Cocks, A. C. F., and Hayhurst, D. R. "The Elastic Response of Creep Damaged Materials," *J. Appl. Mech.*, Vol. 56, 1989, pp. 493–498.

28 Kim, K. T. and Suh, J. "Fracture of Alumina Tube under Combined Tension/Torsion," *J. Am. Ceram. Soc.*, Vol. 75, No. 4, 1992, pp. 896–902.

PART 3

NONDESTRUCTIVE EVALUATION

PART

NONDESTRUCTIVE EVALUATION

Acoustic Microscopic Characterization of Fiber/Matrix Interface of SiC Fiber-Reinforced Reaction-Bonded Si_3N_4 Matrix Composites

SHAMACHARY SATHISH,[1] WILLIAM T. YOST,[2]
JOHN H. CANTRELL,[2] EDWARD R. GENERAZIO,[3]
RAMAKRISHNA T. BHATT[4] and JEFFREY I. ELDRIDGE[4]

OVERVIEW

SCANNING Acoustic Microscopic images of SiC fibers (Textron SCS-6) in reaction-bonded Si_3N_4 matrix have been obtained at a frequency of 200 MHz. The same specimens have been investigated using a fiber push-out technique to determine the interfacial shear strength. Systematic correlation between interfacial shear strength and acoustic images has been observed. This suggests that SAM can be used as a possible nondestructive technique to evaluate interfacial shear strength in fiber-reinforced ceramic matrix composites.

INTRODUCTION

The strength and toughness of fiber composites depend, to a large extent, on the interfacial shear strength between the fiber and matrix. This depends on the chemical nature and physical features of the surfaces. A determination of this strength is extremely important in predicting mechanical performance of the composite materials.

In general, the interfacial shear strength (ISS) in a fiber-reinforced composite can be determined by any of the four destructive techniques: 1) matrix crack spacing method, 2) fiber push-out method, 3) fiber push-in

---

[1]Analytical Services & Materials Inc., Hampton, VA, U.S.A.
[2]NASA Langley Research Center, Hampton, VA, U.S.A.
[3]NASA Lewis Research Center, Cleveland, OH, U.S.A.; current address NASA Langley Research Center, Hampton, VA, U.S.A.
[4]NASA Lewis Research Center, Cleveland, OH, U.S.A.

method, and 4) single fiber pull-out method. These techniques measure the degree of chemical bonding or the interfacial frictional strength or both. Theoretical discussions of these techniques have been described in Reference [1]. Recently, an ultrasonic technique [2] has been proposed to study the interfacial shear strength in ceramic fiber-reinforced ceramic matrix composites. In this chapter, we present a nondestructive technique to examine the interfacial shear strength in a qualitative way using the imaging capabilities of a scanning acoustic microscope (SAM).

The SAM is a powerful tool to image both surface and near-surface microstructural features in materials. These images reflect the variation of elastic properties of the materials. The contrast in an image obtained when the acoustic waves are focused on the top surface of a specimen can be directly related to the variation of acoustic impedance of the material. When the microscope is operated at a negative defocus, subsurface defects in the material can be observed. This mode has been useful in the observation of delaminations, thin films [3], and interphasial structure [4]. Whenever the acoustic lens generates surface acoustic waves on the specimen and receives these waves, then the contrast in the image is dominated by the propagation characteristics of these waves. It is one of the most sensitive modes of operation of the SAM. In the early developmental stages, SAM was used to examine the interface between thin films and substrates. Although the debonded and bonded areas could be distinguished, it was not possible to estimate the strength of adhesion [5]. An attempt to relate the surface wave velocity variations to the adhesion has been made with limited success [6]. Fiber composites have been extensively studied using SAM, especially to study the delamination and surface and subsurface defects [7]. Fiber/matrix interfaces have been investigated, and useful information about the generation of micro cracks and growth of interphasial zone has been obtained [5]. It has never been used, however, to infer the interfacial shear strength in a fiber composite.

In our approach to characterize the fiber/matrix interface, the fibers in the matrix are imaged edge on at a known negative defocus. In this way, it is possible to examine individual fibers as in the case of the fiber push-out or the fiber push-in techniques. Since the fibers and matrix are both ceramics, it is easy to generate Rayleigh surface waves on both. These waves are sensitive to a loss of contact between the fiber and matrix. The amplitude of the Rayleigh waves reflected at the interface changes the contrast in the images of the fiber, depending on the interfacial shear strength between the fiber and the matrix.

EXPERIMENTAL TECHNIQUE

Two specimens of reaction-bonded Si_3N_4 (RBSN) with unidirectional silicon carbide fibers (SCS-6) were used in this investigation. The interfacial shear strength of the two specimens was measured by fiber push-out technique. Both specimens had nearly the same density and fiber loading (±24 vol%) but different interfacial shear strengths (high 13.1 MPa and low <5 MPa). Prior to the fiber push-out tests, the samples were polished with 1.0-μm diamond grit to obtain a flat surface.

The specimens were imaged using an acoustic microscope utilizing a lens operating at a frequency of 200 MHz. A drop of water was used between the lens and sample as a coupling fluid. Several fibers in both specimens were imaged, keeping all the parameters of the microscope identical to obtain a good comparison of the images.

RESULTS AND DISCUSSIONS

Acoustic images obtained at a defocus of $z = -16$ μm for both specimens are shown in Figures 7.1a and 7.2a. A diametrical line scan of the images is shown in Figures 7.1b and 7.2b. At this frequency, the two-point resolution of the SAM is about 8 μm. The dominant features of the images are the fringes observed inside the fiber itself. Even here, some of the fringes are very strong compared to others. To understand the origin of these fringes, it is important to know the microstructure of the fiber itself. The microstructural examination of the fibers with high resolution TEM

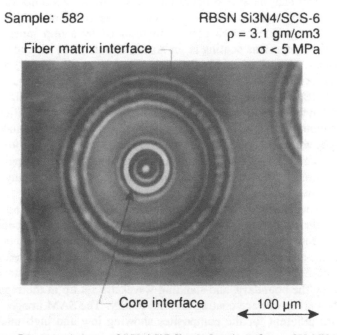

Sample: 582   RBSN Si3N4/SCS-6
ρ = 3.1 gm/cm3
Fiber matrix interface σ < 5 MPa

Core interface   100 μm

Figure 7.1a Acoustic image of SCS-6 SiC fiber in Sample A (freq = 200 MHz).

Standing waves due to weak bond at fiber to
matrix interface

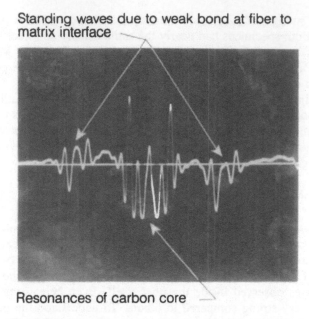

Resonances of carbon core

Figure 7.1b Line scan across the diameter of the fiber in Sample A.

[8] has shown that the SCS-6 fibers themselves are microcomposites consisting of a carbon core, two carbon-rich SiC regions, and a carbon-rich surface coating. The carbon core is surrounded by a thin inner carbon coating of 1.5 μm. This coating is, in turn, surrounded by silicon carbide. The carbon-rich surface coating consists of two layers, each 1.5 μm thick.

As we move from the center of the fiber to the interface between the SiC fiber and RBSN matrix, the fringe pattern changes. The Rayleigh waves generated in the carbon core are reflected back and forth by the silicon carbide surrounding it. The impedance mismatch between carbon and SiC is quite high, which is the reason for the large reflectivity and the formation of strong standing waves in the interior of carbon core. A comparison of the images in this region of the two specimens does not show any appreciable differences.

As we move further toward the interface between the SiC and RBSN matrix, a clear difference is observed in the images of the two specimens. This is the most interesting feature for the characterization of the interface between the fiber and the matrix. The Rayleigh waves generated on the surface of SiC fibers propagate toward the interface. The waves are reflected by the boundary, and standing waves are set up in the region between the carbon surface coating and the RBSN. The SAM images of this region are different for the composites showing low and high interfacial shear strength. For composites showing low interfacial shear strength, the fringe pattern and the peaks in the line scan are very strong. On the other hand, for the composites showing high interfacial shear strength, the fringes are smeared out and the line scan also shows blurred features.

To understand these features, it is necessary to examine the nature of the reflection of the Rayleigh waves at the boundary. At the interface, part of the energy of the Rayleigh waves is reflected, and part is transmitted into the RBSN matrix. The boundary conditions at the interface and the difference in the Rayleigh wave acoustic impedance [$Z_R = \varrho V_R$, where V_R is the Rayleigh wave velocity] between the two materials determine the amount of energy reflected or transmitted.

$$R_{ray} = \frac{Z_{R(SiC)} - Z_{R(Si_3N_4)}}{Z_{R(SiC)} + Z_{R(Si_3N_4)}} \tag{1}$$

The Rayleigh wave acoustic impedance for the two ceramics is calculated from the density and the sound velocities given in Reference [9]. The Z_R for the two ceramics [$Z_{R(SiC)} = 21.85$ Mrayl, $Z_{R(Si_3N_4)} = 18.2$ Mrayl] are close to each other. With these parameters in mind, let us look at the

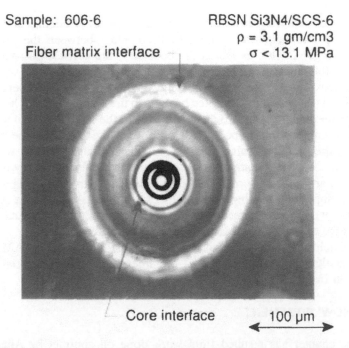

Sample: 606-6

Fiber matrix interface

RBSN Si3N4/SCS-6
ρ = 3.1 gm/cm3
σ < 13.1 MPa

Core interface   100 μm

Figure 7.2a Acoustic image of SCS-6 SiC fiber in Sample B (freq = 200 MHz).

Damped standing waves due to strong bond at
fiber to matrix interface

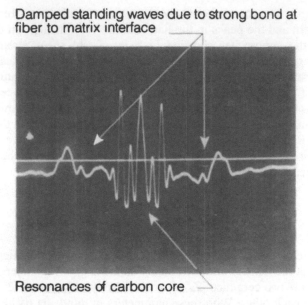

Resonances of carbon core

Figure 7.2b Line scan across the diameter of the fiber in Sample B.

composites exhibiting two different interfacial shear strengths. In the case of strong interfacial shear strength, the contact between the fiber and matrix is good. This means, at the boundary, the particle displacements for both longitudinal and transverse components of the Rayleigh waves are continuous. In this case, approximately 10 percent of the total energy incident from the SiC side is reflected back from the boundary [5]. The amplitude of the standing waves is thus quite low. In the case of weak interfacial shear strength, there is poor contact between the fiber and the matrix. Hence, the particle displacements across the boundary will be weak. In particular, the shear component is weak and the reflected energy is much higher than in the case of strong interfacial shear strength. This leads to higher amplitude standing waves in the SiC region and to fringe patterns that are in qualitative agreement with the images and the line scans for the two cases of strong and weak interfacial shear strength.

In conclusion, we have demonstrated the feasibility of using SAM to obtain qualitative information about the fiber to matrix interfacial shear strength in ceramic composites.

ACKNOWLEDGEMENT

This chapter has resulted from work done on contract by Analytical Services & Materials Inc., NAS1-19236.

REFERENCES

1 Kerans, R. J., Hay, R. S., Pagano, N. J., and Partasarathy, T. A. "The Role of the Fiber-Matrix Interface in Ceramic Composites," *Ceram. Bull.*, Vol. 68, No. 4, 1989, pp. 429–440.

2 Generazio, Edward R. "Nondestructive Evaluation of Ceramic and Metal Matrix Composites for NASA's HITEMP and Enabling Propulsion Materials Programs," NASA TM-105807, August 1992.

3 Bray, R. C., Quate, C. F., Calhoun, J., and Koch, R. "Film Adhesion Studies with the Acoustic Microscope," *Thin. Solid Films.*, Vol. 74, 1980, pp. 295–302.

4 Cognard, J., Sathish, S., Kulik, A., and Gremaud, G. "Scanning Acoustic Microscopy of the Cellular Structure of the Interphase in a Metal-Adhesive Bond," *J. Adhesion*, Vol. 32, 1990, pp. 45–49.

5 Briggs, G. A. D. *Acoustic Microscopy.* Oxford Univ. Press, Oxford, New York, 1992.

6 Addison, R. C., Jr., Somekh, M., and Briggs, G. A. D. "Techniques for the Characterization of Film Adhesion," *IEEE Ultrsonics. Symposium*, 1986, pp. 775–782.

7 Hollis, R. L., Hammer, R., and Al-Jaroudi, M. Y. "Subsurface Imaging of Glass Fibers in Polycarbonate Composite by Acoustic Microscopy," *J. Mater. Sci.*, Vol. 19, 1984, pp. 1897–1903.

8 Ning, X. J. and Pirouz, P. "The Microstructure of SCS-6 SiC Fiber," *J. Mater. Res.*, Vol. 6, No. 10, 1991, pp. 2234–2248.

9 Roth, D. J., Stang, D. B., Swickard, S. M., and DeGuire, M. R. "Review and Statistical Analysis of the Ultrasonic Velocity Method for Estimating the Porosity Fraction in Polycrystalline Materials," NASA TM-102501, March 1990.

CHAPTER 8

Scanning Acoustic Microscopy of SCS-6 SiC Fibers in Titanium Matrices

SHAMACHARY SATHISH,[1] WILLIAM T. YOST,[2]
JOHN H. CANTRELL,[2] EDWARD R. GENERAZIO,[3]
REBECCA A. MACKAY[4] and KAREN M. B. TAMINGER[2]

OVERVIEW

IMAGES of SiC fibers (Textron SCS-6) using scanning acoustic micros-copy have been obtained at a frequency of 1 GHz. The contrast observed in different regions of the fiber is explained in terms of variations in the Young's modulus and mass density of the fiber. For the first time, elastic property variations in the carbon-rich region of the fiber is reported.

INTRODUCTION

In high-performance metal matrix and ceramic matrix composites, sili-con carbide fibers are used as reinforcement materials. These fibers have a high tensile modulus and low mass density. They can also withstand very high temperatures and, hence, are suitable for high-temperature applica-tions. The overall physical and mechanical properties of the composites depend on the microstructure of the materials involved. The silicon car-bide fibers generally used for reinforcement have diameters in the range of 100–200 μm and exhibit certain microstructural characteristics that strongly influence the properties of the composite. The microstructure of SCS-6 SiC fibers has been investigated in detail by Ning and Pirouz [1,2] using scanning electron microscopy (SEM), transmission electron micros-copy (TEM), and high-resolution electron microscopy (HREM). Their

[1]Analytical Services & Materials, Inc., Hampton, VA, U.S.A.
[2]NASA Langley Research Center, Hampton, VA, U.S.A.
[3]NASA Lewis Research Center, Cleveland, OH, U.S.A.; current address NASA Langley Research Center, Hampton, VA, U.S.A.
[4]NASA Lewis Research Center, Cleveland, OH, U.S.A.

studies clearly reveal that the fiber itself is a microcomposite consisting of several layers of carbon and silicon carbide having distinct microstructural features. They have also mapped the variation of the chemical composition in different regions of the fiber using scanning auger microscopy, electron energy loss spectroscopy (EELS), and parallel electron energy loss spectroscopy (PEELS).

Since the microstructure plays a dominant role in the determination of the mechanical properties of the material, it is important to characterize the elastic properties of the different regions of the fiber. Although the above-mentioned electron microscopies and spectroscopies are indispensable in revealing the microstructure and chemical composition of the fiber, they do not provide information about the mechanical properties. In this chapter, we present an investigation of the SCS-6 fiber using a scanning acoustic microscope (SAM), which allows an evaluation of the elastic properties on a microscopic scale on the order of $10-15$ μm. A SAM image of an SCS-6 fiber is presented, and the contrast in the image is explained on the basis of the microstructure and chemical analysis determined by Ning and Pirouz [1]. An estimate of the Young's modulus of different regions of the fiber is also presented. It is, to the authors' knowledge, the first time that an in situ elastic modulus measurement over microscopic dimensions has been reported for any material.

The SAM is proving to be a powerful tool for the nondestructive characterization of materials at the microstructural level. The contrast in a SAM image is directly related to the elastic properties of the material. The nature and theory of the contrast has been explored in detail in several articles in the literature [3,4]. In brief, when the acoustic lens is positioned to focus the generated acoustic wave to a point on the specimen surface, the amplitude of the reflected acoustic signal is dependent upon the value of the acoustic reflectivity R at that point in the material. R depends on the density and elastic modulus of the material at that point. If the acoustic reflectivity varies in the material from point to point, the amplitude of the reflected signal will vary accordingly and lead to contrast in the generated image. If the acoustic lens is positioned such that the generated acoustic signal is brought to focus at a point below the specimen surface, a surface acoustic wave is generated under certain conditions, and the contrast is then dominated by the interference between the surface acoustic wave and the wave specularly reflected at the specimen surface. Detailed discussions of operating in this latter mode to assess the elastic properties of materials can be found elsewhere [5].

EXPERIMENTAL TECHNIQUE

The composite specimen chosen for investigation consists of SCS-6 fibers embedded in a Ti(15-3) matrix. The specimen was polished with a 0.25-μm diamond grit paste to reduce the surface roughness, which otherwise would dominate the SAM image contrast. Because of the great differences in the hardness of the materials comprising the composite, some remnant topography contributing to the image contrast likely still exists. However, the remnant topography is not expected to alter the conclusions of the present study. The specimen was imaged in a scanning acoustic microscope operating at a frequency of 1.0 GHz. Distilled water was used for the coupling fluid between the lens and the specimen.

A SAM image of an embedded fiber is shown in Figure 8.1. A line scan across the horizontal diameter of the fiber is shown in Figure 8.2 to illustrate the variation in the reflected signal amplitude in different regions of the fiber. Every point on this line scan can be expressed as [3,6]

$$f(Y,\sigma,\varrho) = \int_0^\vartheta R(\vartheta)P(\vartheta)e^{-2ikz} \cos\vartheta \cos\vartheta \sin\vartheta d\vartheta$$

where $R(\vartheta)$ is the reflectivity for a liquid-solid interface and $P(\vartheta)$ is the pupil function of the lens, ϑ is the opening angle of the lens, z is the distance between the lens and the sample measured with respect to focus, and

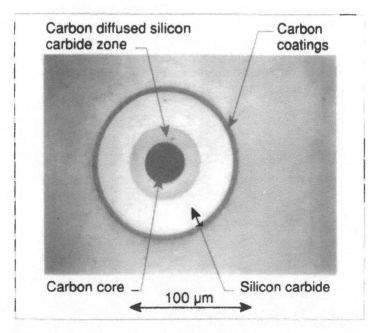

Figure 8.1 Acoustic image of SCS-6 SiC fiber in Ti(15-3) matrix (frequency = 1.0 GHz).

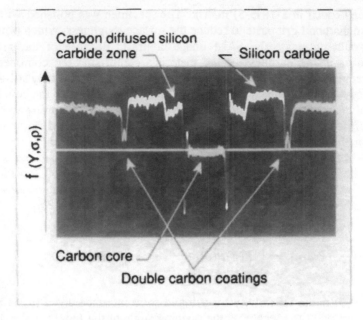

Figure 8.2 Line scan across the diameter of the fiber.

k is the wavevector in the coupling fluid. The equation above is also used to express $V(z)$. Here, we are denoting it as $f(Y,\sigma,\varrho)$ because it is obtained at a given z and to show that it depends on the density ϱ_s, the Young's modulus Y, and Poisson's ratio σ of the material at a given point.

For an isotropic solid-liquid interface, the reflection coefficient $R(\vartheta)$ is given in terms of acoustic impedances, Z_{tot} and Z as

$$R(\vartheta) = \frac{Z_{tot} - Z}{Z_{tot} + Z}$$

where

$$Z_{tot} = \frac{\sqrt{\varrho_w^3}}{\varrho_s}\frac{\sqrt{Y}}{K^3}\left[\sqrt{m}\sqrt{\frac{A}{B}} + 4Y\sqrt{n^3}\right]$$

$$A = \frac{\varrho_s}{\varrho_w}\sqrt{K\varrho_w} - nY\sin\vartheta^2$$

$$B = \frac{\varrho_s}{\varrho_w}\sqrt{K\varrho_w} - mY\sin\vartheta^2$$

$$m = \frac{(1 - \sigma)}{(1 + \sigma)(1 - 2\sigma)} \text{ and } n = \frac{1}{2(1 + \sigma)}$$

$$Z = \frac{\sqrt{K\varrho_w}}{\cos \vartheta}$$

ϱ_w is the density of the liquid, and K is its bulk modulus.

RESULTS AND DISCUSSIONS

The image (Figure 8.1) is obtained by positioning the acoustic lens such that the generated acoustic wave is focused to a point only slightly below the specimen surface, so that an acoustic surface wave is generated with minimal opportunity for interaction with the specularly reflected signal. There is also a very slight tilt in the specimen, which promotes a small variation in the reflected signal amplitude observed as we move from left to right across the image. A few faint fringes are also observed in the image, which are explained later in the chapter.

According to the HREM and the electron diffraction observations of Ning and Pirouz [1], the SCS-6 fiber has the following microstructure: the central or core portion of the fiber is a filament consisting of turbostratic carbon (TC) blocks having dimensions in the range 1–50 nm; the c-axis of these blocks are randomly oriented throughout the core region. Surrounding the core region is a 1.5-μm thick coating of pyrolitic carbon consisting of TC blocks with dimensions in the range 30–50 nm, which are arranged with the c-axis of the blocks preferentially (but not exclusively) aligned along the direction radial to the carbon filament axis. This inner coating of pyrolitic carbon thus has microtexture. Immediately outside the pyrolitic carbon layer are four layers of silicon carbide. The layer SiC-1 adjacent to the inner carbon coating consists of rod-like grains of β-SiC. Nearest the inner carbon coating, the β-SiC grains are roughly 5–15 nm in length and are randomly oriented. At larger distances from the inner carbon coating, the grains become larger (of the order 50–150 nm in length) and become increasingly aligned along the fiber radial direction. Adjacent to the SiC-1 layer is a SiC-2 layer with a pronounced twin orientation between the grains. The SiC-3 layer adjacent to the SiC-2 layer is a region of heavily faulted grains with the faulting increasing as the grain size increases. The SiC-4 layer, starting at approximately 15 μm from the core region, is clearly demarcated from the SiC-3 layer by the appearance of grains that are roughly twice the size of the grains in the SiC-3 layer. Chemical analysis suggests [1] that a surplus of carbon resides in the SiC grain boundaries of the SiC-1, 2, and 3-layers, but the SiC-4 layer has no such surplus. At the extreme edge of the fiber is a coated layer consisting of SiC particles embedded in a carbon matrix (referred to as the outer carbonaceous layer).

In SAM images, the above microstructural features of the SCS-6 fiber are explained by looking at both the image and the line scan through the diameter of a fiber. The contrast in SAM is due to the variation in the elastic properties and in the mass densities in the different regions of the specimen. When the image is taken close to the focal point of the acoustic lens, the bright and dark shades in the image are directly related to the variations in the acoustic reflection coefficient of the region. The reflection coefficient of SiC along random crystalline directions is, on average, higher than that of carbon along random crystalline directions. Hence, the SiC regions are generally more highly reflective than the carbon regions under near focus imaging conditions. The TC blocks in the core of the fiber are randomly oriented, while in both the inner and outer carbonaceous layers, the TC blocks have a preferred orientation. Turbostratic carbon is elastically highly anisotropic and the elastic moduli most sensitive to inter-basal plane disorder (i.e., C_{33} and C_{44} in Voigt notation) are quite low [7]. These elastic moduli dominate the elastic properties along the c-axis of the turbostratic carbon and may be responsible for the dramatic decrease in the reflectivity, hence signal level in the SAM image, as the result of texture in both the inner and outer carbonaceous layers. The line scan of Figure 8.2 clearly shows a large dip in the amplitude in the carbon coating compared to the core. Optical microscopic and SEM observations of this layer are not as easily obtained as in SAM. Both the SAM image and the line scan of the outer carbon layer indicate a higher reflectivity than the inner carbon coating. This is perhaps explained by taking into account the SiC particles embedded in the carbonaceous matrix of the outer coating.

Ning and Pirouz [1] have shown that, just after the inner carbon coating, there are three layers of SiC with different structures. The carbon content varies radially in these regions. The contrast even in the HRTEM between different layers is very small. In SAM, it is not possible to observe any changes in the contrast at all. One of the reasons might be that these different regions have similar crystalline orientation and, hence, have the same reflection coefficient. SAM sees it as a single layer of SiC. Just after this region, there is stoichiometric SiC layer. This layer must have a higher reflection coefficient than the inner SiC layer because of its higher purity. This can be clearly observed in the line scan. The inner layer of SiC has a lower reflectivity compared to the stoichiometric SiC. A careful observation of the line scan in the inner SiC layer shows a continuously varying reflectivity, which is an indication of the variation of carbon content and, hence, the elastic moduli and density.

A simple quantitative estimation of the approximate shear modulus can be performed by using the images obtained on these fibers. When the lens is slightly defocused during imaging in these fibers, fringe patterns are observed. These fringes are due to the propagation of surface acoustic waves at the interface between the specimen and water. These waves are extremely sensitive to the surface discontinuities and defects. When these waves meet these kinds of inhomogenities, they will be reflected and also scattered. Because of this reflection, there could be a formation of standing waves. These are seen as fringes in the images. The spacing between the fringes is half the wavelength of the surface waves in the material. The Rayleigh wave velocity in an isotropic material is a few percent lower than the bulk shear wave velocity, and hence, we can approximately determine the shear wave velocity and the shear modulus.

We have observed the Rayleigh wave fringes in different regions of the fiber and evaluated the Rayleigh wave velocity of that particular region. The Rayleigh wave velocity in the central carbon core is 2700 m/s. This is in good agreement with the estimated Rayleigh wave velocity for amorphous carbon using the velocities determined by Reference [8]. In the stoichiometric SiC region, it is 6800 m/s, which is in good agreement with the estimates of the Rayleigh wave velocity from the measured sound velocities [8]. The SiC in this region can be assumed to be elastically isotropic, and hence, a Poisson's ratio $\sigma = 0.19$ can be taken. This gives a Young's modulus of 427 GPa. In the region between the stoichiometric SiC and the inner carbon coating, the average velocity is 5900 m/s. In fact, in this region, it is known that the carbon content varies across the radial direction. It would be extremely important to determine the variation in moduli and density point by point. At the moment, it is difficult to determine them, and hence, only an average over the whole region can be estimated. Assuming that the Poisson's ratio for this region is between that of SiC and carbon, the Young's modulus is in the range of 105–109 GPa.

In conclusion, it is clear from our observations that the SCS-6 silicon carbide fiber itself is a microcomposite consisting of materials with different mechanical properties.

ACKNOWLEDGEMENT

This chapter has resulted from work done on contract by Analytical Services & Materials Inc., NAS1-19236.

REFERENCES

1 Ning, X. J. and Pirouz, P. "The Microstructure of SCS-6 SiC Fiber," *J. Mater. Res.*, Vol. 6, No. 10, 1991, pp. 2234–2248.

2 Ning, X. J., Pirouz, P., Lagerlof, K. P. D., and DiCarlo, J. "The Structure of Carbon in Chemically Vapor Deposited SiC Monofilaments," *J. Mater. Res.*, Vol. 5, No. 12, 1990, pp. 2865–2876.

3 Atalar, A. "An Angular-Spectrum Approach to Contrast in Reflection Acoustic Microscopy," *J. Appl. Phys.*, Vol. 49, No. 10, 1978, pp. 5130–5139.

4 Wickramasinghe, H. K. "Contrast and Imaging Performance in the Scanning Acoustic Microscope," *J. Appl. Phys.*, Vol. 50, No. 2, 1979, pp. 664–672.

5 Kushibiki, J. I. and Chubachi, N. "Material Characterization by Line-Focus-Beam Acoustic Microscope," *IEEE Trans. on Sonics and Ultrasonics*, Vol. SU-32, No. 2, 1985, pp. 189–212.

6 Sheppard, C. J. R. and Wilson, T. "Effects of High Angles of Convergence on V(Z) in the Scanning Acoustic Microscope," *Appl. Phys. Lett.*, Vol. 38, No. 11, 1981, pp. 858–859.

7 Dresselhaus, M. S., Dresselhaus, G., Sugihara, K., Spain, I. L., and Goldberg, H. A. *Graphite Fibers and Filaments*. Springer, Berlin, Heidelberg, 1988.

8 Selfridge, A. R. "Approximate Material Properties in Isotropic Materials," *IEEE Trans. on Sonics and Ultrasonics*, Vol. SU-32, No. 3, 1985, pp. 381–394.

CHAPTER 9

The Use of Local Computerized Tomographic Inspection for Nondestructive Evaluation of Ceramic Components

E. A. SIVERS,[1] W. A. ELLINGSON[2] and D. A. HOLLOWAY[3]

OVERVIEW

X-RAY computerized tomography (CT) has become an important tool for the nondestructive inspection of advanced ceramic components. Recently, local tomography—a variation of conventional (global) CT—has been found to be advantageous for some applications. Unlike global CT, which produces two-dimensional, cross-sectional images of electron density, local CT produces edge-enhanced versions of these images. At each density interface, an "overshoot" (a rapid increase/decrease in CT number, followed immediately by a decrease/increase of equal magnitude) appears that is proportional to the magnitude of the density difference. Because only the contrast between adjacent constituents in an image is displayed, local CT can improve the detectability of features in an object having a wide range of densities. It is also useful for producing high-resolution images of a region of interest (ROI) in a component too large to be encompassed by the CT X-ray beam. Without a full data set, global CT images are subject to extreme density shading. Time-consuming methods introduced previously to remedy these global density artifacts require specialized processing to replace or approximate the missing data outside the desired volume. However, because local CT operates upon only a narrow range of data, it is insensitive to missing data and produces a flat ROI image with good contrast discrimination. For the same reason, local CT reconstructions require less computation time than global CT reconstructions. It is also possible to use local CT images for dimensional analysis

[1]Argonne National Laboratory, Lake Oswego, OR, U.S.A.
[2]Argonne National Laboratory, Argonne, IL, U.S.A.
[3]Insite Industries, Durham, NC, U.S.A.

by performing cross correlation with a mask that duplicates the overshoot. Thus, local CT is a valuable new X-ray inspection technique for positional information.

After a brief explanation of the differences in implementation between global and local CT reconstructions, we compare the contrast, noise, and resolution in both types of image. Using simulated and real data, we demonstrate that contrast detectability in local CT images seen on a practical display is often superior to that in global CT images. It will also be shown that, although noise is somewhat higher in local CT reconstructions, it often appears to be amplified further because of the relative scaling between local and global images. Finally, several images of green Si_3N_4 pressure-slip-cast parts taken by a 3-D X-ray CT system will be used to compare global and local CT for ROI scans.

CT THEORY

A parallel-beam form of local CT was published by Vainberg et al. [1] in 1981. A mathematically comprehensive version was introduced independently by Smith and Keinert [2] in 1985. Additional contributions to the practical implementation of the method have been made by Faridani, Keinert, et al. [3] and by Faridani, Ritman, et al. [4] Although it is not appropriate to describe the theory in great detail here, some mention should be made of the differences between global and local CT. The presentation of the radon inversion formula for global CT and the analogous inversion formula for local CT serves to illustrate the two methods. Although this discussion will be limited to two-dimensional CT, it can be extended to three-dimensional CT using the method of Feldkamp et al. [5]

Ideally, a CT reconstruction produces an image of the linear X-ray attenuation coefficient $\mu(r, \theta)$ in a plane through the object being scanned. For monoenergetic X-rays, the signal I produced at a detector by an X-ray beam of intensity I_o that has traversed a path length x of a material having linear attenuation coefficient μ is given by

$$I = I_o e^{-\mu x} \tag{1}$$

Dividing both sides of Equation (1) by I_o and taking the natural logarithm produces CT projection data P, which is used by the reconstruction process to produce an image of $\mu(r, \theta)$:

$$P = \ln\left(\frac{I_o}{I}\right) = \mu x \tag{2}$$

In reality, the global CT reconstruction process is equivalent to convolving the ideal image function $\mu(r, \theta)$ with a two-dimensional point-spread function (PSF). The PSF that is characteristic of a real imaging system contains contributions from the X-ray beam width and sample spacing, as well as the CT mathematical inversion process. Because the physical contributions are inherent in a given imaging system, variables in the inversion process are usually chosen to complement them. In addition, statistical and electrical signal fluctuations in the data often necessitate smoothing the PSF to reduce image noise. This discussion demonstrates how the CT inversion formulas are related to the PSF.

Global CT reconstruction produces an image that is the two-dimensional convolution of the PSF $e(r, \theta)$ with the ideal image function $\mu(r, \theta)$. Equations (3)–(8) describe the global fan-beam CT reconstruction algorithm. Here, $p_{\parallel}[e](r, \theta)$ is the parallel-beam, 1-D projection of the PSF. The PSF is usually assumed to possess circular symmetry so that we need discuss only its one-dimensional projection. $P(\alpha_\bullet)$ is the one-dimensional, fan-beam projection of the object, i.e., the CT data; ϕ is the source rotation angle variable; α is the fan angle variable; $2A$ is the fan angle; and D is the source-to-object distance. These parameters are illustrated in Figure 9.1 for a lathe-bed CT scanner. Equation (3) defines the reconstructed function in terms of an integral over the convolved projection data $C[P(\alpha_\bullet)](r, \theta)$, the operation of which is defined in Equations (4) and (5).

$$e(r, \theta) * \mu(r, \theta) = \frac{1}{4\pi^2} \int_0^{2\pi} \frac{D}{(D^2 + r^2 - 2Dr \cos(\theta - \phi))}$$

$$\times \ C[P(\alpha_\bullet)](r, \theta)d\phi \tag{3}$$

$$C[P(\alpha_\bullet)](r, \theta) = \int_{-A}^{A} \cos \alpha_\bullet' \, P(\alpha_\bullet') \, G(\sin(\alpha_\bullet(r, \theta) - \alpha_\bullet'))d\alpha_\bullet' \tag{4}$$

$$\alpha_\bullet(r, \theta) = \tan^{-1}\left(\frac{r \sin(\phi - \theta)}{D - r \cos(\phi - \theta)}\right) \tag{5}$$

Here, $G(\sin(\alpha))$ is the global convolution filter, which can be approximated by a constant times the Lambda operator applied to the parallel projection of the PSF.

$$G(\sin(\alpha)) \cong \text{const } \Lambda[p_{\parallel}[e]](\sin(\alpha)) \tag{6}$$

The Lambda operator is defined by Equations (7) and (8), where s is the

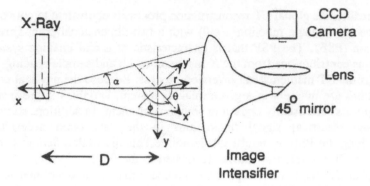

Figure 9.1 Illustration of coordinates and components of lathe-bed X-ray computerized tomographic scanner.

dimension of the space (2 in this case) and the symbol $*$ denotes convolution.

$$\Lambda[u](x) = (n - 1)B(n, 1) \sum_{j=1}^{s} \left(\frac{x_j}{|x|^{n-1}}\right) * \frac{\partial u}{\partial x_j} \tag{7}$$

$$B(n, \gamma) = \frac{\Gamma\left(\frac{(n - \gamma)}{2}\right)}{2^\gamma \pi^{n/2} \Gamma\left(\frac{\gamma}{2}\right)} \tag{8}$$

Alternatively, Γ can be defined in terms of the Fourier transform by the following equation where ξ is the frequency variable $1/l$.

$$F[\Lambda[l]](\xi) = |\xi| F[l](\xi) \tag{9}$$

Equations (3)–(8) must be discretized for implementation on a digital computer. An excellent discussion of the digital implementation of the global CT inversion process is found in Ramachandran and Lakshminarayanan [6]. However, the global convolution filter introduced there accentuates statistical noise. In the global reconstructions presented here, the smoothed global convolution filter of Shepp and Logan [7] is used.

The local CT reconstruction process is very similar to the global process except that the resulting image is the two-dimensional convolution of the point-spread function $e(r, \theta)$ with $\Lambda[\mu(r, \theta)]$. Equations (9)–(12) describe the local fan-beam CT reconstruction algorithm. Essentially, local CT is obtained by the operation of Λ on Equation (3). However, with Equation (9) it is possible to demonstrate that $\Lambda[e]*\mu = e*\Lambda[\mu]$. Thus, the local reconstruction is a good approximation to $\Lambda\mu$, which is different from μ quantitatively but has the same singularities and details. Equation (10) is analogous to Equation (3). Use has been made of the fact that $\Lambda^2 = -\nabla^2$, i.e., the square of the Lambda function is the negative of the Laplacian. Equation (10) defines the local reconstruction of a function in terms of an integral over the convolved projection data $K[P(\alpha_*)](r,\theta)$, the operation of which is defined in Equation (11).

$$e(r,\theta) * \Lambda[\mu(r,\theta)] = \frac{1}{4\pi^2} \int_0^{2\pi} \frac{D}{[D^2 + r^2 - 2Dr\cos(\theta - \phi)]^{3/2}}$$

$$\times\ K[P(\alpha_*)](r,\theta)d\phi \tag{10}$$

$$K[P(\alpha_*)](r,\theta) = \int_{-A}^{A} \cos\alpha_*' P(\alpha_*')L(\sin(\alpha_*(r,\theta) - \alpha_*'))d\alpha_*' \tag{11}$$

Here, $L(\sin(\alpha))$ is the local convolution filter, which can be approximated by a constant times the Laplacian operator applied to the parallel projection of the point-spread function.

$$L(\sin(\alpha)) \cong -\text{const}\ \nabla^2[p_{||}[e]](\sin(\alpha)) \tag{12}$$

The basic operational differences between global and local CT are the extra power of the factor in the denominator of Equations (3) and (10) and the form of the convolution filter.

The point-spread functions (PSFs) used for global and local CT are also different, but this is not fundamental. The functional forms of the PSFs are chosen to facilitate the mathematical reconstruction process. Also, these PSFs are assumed to have circular symmetry so that the projection is not dependent upon the variable θ. Equation (13) defines the normalized PSF $e_m(\eta)$ used to construct a local CT filter [3]; m is a variable parameter. The dimensionless variable η is the quotient of the spatial variable r divided by the radius of the PSF ϱ, $(\eta = r/\varrho)$, and has a maximum value of 1.0.

$$e_m(\eta) = \frac{1}{2\pi}(2m + 3)(1 - |\eta|)^{m+1/2}; \quad |\eta| \le 1 \tag{13}$$

Applying Equation (12) and normalizing to eliminate the constant produces the local convolution filter defined by

$$L_m(\eta) = \pi l_m (1 - \eta^2)^{m-1} (1 - (2m + 1)\eta^2) \qquad (14)$$

$$l_m = \frac{\Gamma\left(m + \frac{5}{2}\right)}{2\pi^{5/2}m!} \qquad (15)$$

For digital implementation, the local filter must be discretized at a rate consistent with the physical sampling of the CT scanner. Nyquist sampling theory requires at least two samples per PSF, but the actual PSF width varies radially within an image made by a fan-beam CT scanner. In practice, the filter is forced to have the same sampling as the data, and the PSF radius is taken to be an integer multiple of the sample spacing; m is chosen such that the sum of the terms in the filter is zero. In these studies, symmetrical local filters having a total of five, nine, and thirteen terms have been used with corresponding m values of 11.4174, 11.703, and 7.7353, respectively.

There is some residual cupping in a local reconstruction of even a full-data set for a scan of a homogeneous object. It is shown in Faridani et al. [4] that this cupping can be corrected by adding a scaled version of a simple back projection $\epsilon\Lambda^{-1} \mu(r,\theta)$ to the reconstruction:

$$\Lambda^{-1}\mu(r,\theta) = \frac{1}{4\pi^2} \int_0^{2\pi} P(\alpha_\phi)d\phi \qquad (16)$$

where ϵ is a constant on the order of 10^{-7} and $\alpha_\phi(r,\theta)$ is given by Equation (5). If a local filter is chosen as described above, this cupping is very small, and the constant ϵ will have slightly different values for different filters.

SCALING AND CONTRAST IN LOCAL CT

Because reconstructed values in global CT images are the linear attenuation coefficients μ of the scanned objects, the choice of a scaling factor and the definition of contrast are unambiguous. The reconstructed global image is multiplied by the constant needed to map the range of attenuation coefficients into a range of about 4096 integer values. These integer values are displayed as shades of gray on a cathode ray tube (CRT). The contrast of a feature having attenuation coefficient μ_f in a constant background having attenuation coefficient μ_b is defined by

$$\text{CONTRAST} = \Delta\mu = |\mu_b - \mu_f| \qquad (17)$$

In a local reconstruction, it is the overshoot at the interface between two materials that is reconstructed and that is proportional to the contrast between them. The reconstructed local image is multiplied by the constant needed to map the range of contrasts onto the range of gray-scale values in a CRT display. Figure 9.2a illustrates the relative ranges of values in unscaled global and local reconstructions of a right circular cylinder. The graph is a superposition of one-dimensional plots of horizontal profiles through the centers of the reconstructed cylinders. The larger, square curve (denoted by *) is the global reconstruction. On this scale, the plots from the local reconstructions are barely visible. Figure 9.2b shows the local reconstructions on a smaller scale. The smallest curve is obtained using the five-point local filter, and the largest is obtained using the thirteen-point filter. The approximate relative multiplicative constants S_n

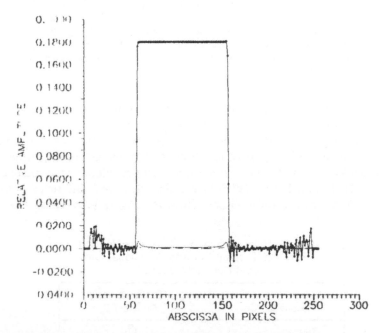

Figure 9.2a Superposition of one-dimensional horizontal profiles through the center of reconstructions of an off-centered, right circular cylinder: ***** global reconstruction; ----- local reconstruction, thirteen-point-filter; ---- local reconstruction, nine-point filter; -- -- local reconstruction, five-point filter.

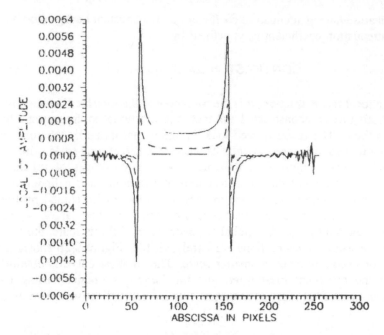

Figure 9.2b Superposition of one-dimensional horizontal profiles through the center of reconstructions of an off-centered, right circular cylinder: ------ local reconstruction, thirteen-point filter; ---- local reconstruction, nine-point filter; -- -- local reconstruction, five-point filter.

needed to map all four reconstructions onto the same scale are shown in Table 9.1. As before, D is the source-to-object distance.

The relative scaling between global and local reconstructions is demonstrated in Figure 9.3. Figure 9.3a compares horizontal profiles through global and local reconstructions of the same right circular cylinder mapped onto a CRT display. For a single object, there is no real difference in the mapping of the two reconstructions. However, Figure 9.3b shows analo-

TABLE 9.1. Relative Scaling Constants Needed
to Equalize Display Contrast for Global and
Local Reconstructions.

Reconstruction	Relative Scaling Constant S_n
Global	1.0
Five-point local	5.53 D
Nine-point local	1.94 D
Thirteen-point local	0.87 D

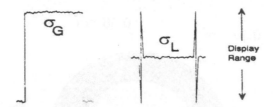

Figure 9.3a One-dimensional profile illustrating equivalent scaling used to map global (left) and local (right) CT reconstructions of a simple right circular cylinder onto a CRT display. Standard deviations of noise on global and local reconstructions are σ_G and σ_L, respectively.

gous plots for the reconstruction of the same cylinder when another cylinder that is twice as attenuating is embedded in it. The scaling of the global reconstruction must be halved to accommodate the display, but the scaling of the local reconstruction remains the same. Thus, the contrast in the displayed global image is effectively half of that in the local image.

This compression of contrast needed to view global CT images would be of little consequence if the human eye could perceive 4096 gray-scale levels in a CRT display. In fact, only about 128 different shades can be distinguished. Thus, to see the entire contrast range, it is necessary to separately map many different parts of the full range onto the 128 shades or to map many adjacent values into one shade to reduce the range. Because viewing all of the possible 1:1 mappings is very time-consuming,

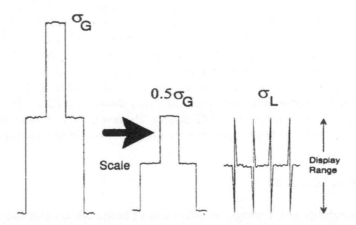

Figure 9.3b One-dimensional profile illustrating equivalent scaling used to map global and local CT reconstruction of a compound right circular cylinder onto a CRT display. The scaling factor 0.5 needed to accommodate the global image makes the global noise σ_G seem smaller than the local noise σ_L.

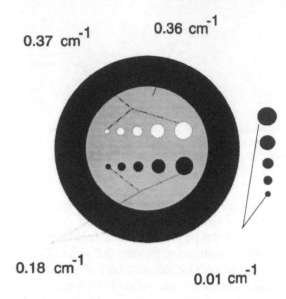

0.37 cm^{-1} 0.36 cm^{-1}

0.18 cm^{-1} 0.01 cm^{-1}

Figure 9.4 Cross section of mathematical test phantom containing cylinders of different X-ray attenuation coefficients.

it is much more common to view the entire range with 128 shades, and this procedure is used here.

This concept is clarified in Figures 9.4 and 9.5. Figure 9.4 shows a simulated test object (phantom) similar to the one described in Figure 9.3b. Linear attenuation coefficients μ are measured in cm^{-1}. Included in the inner cylinder are five cylinders having a contrast of 0.18 cm^{-1} and five cylinders having a contrast of 0.01 cm^{-1}. There are also five cylinders having a contrast of 0.01 cm^{-1} outside the large cylinders. Figure 9.5a shows the global reconstruction of this phantom. In addition to the two large cylinders, only the five high-contrast cylinders are visible. In the corresponding nine-point filter local reconstruction shown in Figure 9.5b, all of the cylinders are visible on a CRT display. However, the five low-contrast cylinders inside the larger cylinders are difficult to see in this photograph.

NOISE IN LOCAL CT

Detectability in CT images is a function of both contrast and statistical noise. As shown in Figure 9.3b, even if the standard deviation of the local statistical noise σ_L is equal to the standard deviation of the global statistical noise σ_G, global noise will appear to be much smaller on the final image if the scaling of the global image is reduced for display. In addition, noise in CT reconstructions is strongly dependent on the convolution filter used.

Figure 9.5a Global CT reconstruction of one plane through test phantom shown in Figure 9.4.

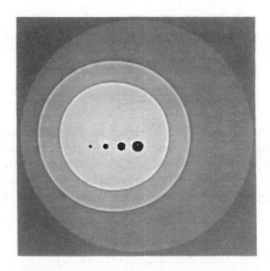

Figure 9.5b Local, nine-point filter, CT reconstruction of one plane through test phantom shown in Figure 9.4.

For comparison, it is possible to compute the noise at the center of the reconstruction of a centered, right circular cylinder by using both reconstruction techniques. For a global CT image, Equation (3) at the center of a centered cylinder reduces to Equation (18):

$$\mu = \sum_{\nu=1}^{V} \Delta\phi\, C(0) \tag{18}$$

$$C(0) = \sum_{j=-(\alpha/\Delta\alpha)}^{\alpha/\Delta\alpha} P(j\Delta\alpha)\, \cos\,(j\Delta\alpha)SL_j \tag{19}$$

where V is the number of discrete X-ray source positions, $\Delta\phi$ is the angular rotation increment of the X-ray source ($\Delta\phi = 2\pi/V$), $C(0)$ is the convolved signal, $P(j\Delta\alpha)$ is the projection data, α is the X-ray fan angle, $\Delta\alpha$ is the X-ray angle increment, and SL_j is the Shepp-Logan [7] convolution filter:

$$SL_j = \frac{1}{\pi^2 x(1 - 4j^2)} \tag{20}$$

Moreover, in practice, the convolved data are linearly interpolated before being summed in Equation (18) (back projection). By the rules of quadrature, the variance of μ will be given by

$$\sigma_G^2 = V \left[\frac{2\pi}{V}\right]^2 \sigma_{C,I}^2 \tag{21}$$

where $\sigma_{C,I}^2$ is the variance of the interpolated, convolved, logged projection data. Because the variance of the logged signal will itself be fairly constant in the center of a "large" cylinder and because the filter K is strongly peaked, cos $(j\Delta\alpha)$ is approximated by 1, and the variance of the convolved, logged signal is given to a good approximation by

$$\sigma_C^2 \cong \frac{\sigma_P^2}{\pi^4 x^2} \left[\sum_{j=-\infty}^{+\infty} \frac{1}{(1 - 4j^2)^2}\right] \tag{22}$$

$$\sigma_C^2 \cong \frac{\sigma_P^2}{8\pi^2 x^2} \tag{23}$$

where σ_P^2 is the common variance of the projection data, which is assumed to be uncorrelated. Equation (23) does not take into account the smoothing of the linear interpolation on the convolved data before back projection. The variance of the interpolated, convolved data is given by

$$\sigma_{\tilde{C},I}^2 = [f^2 + (1 - f)^2]\sigma_{\tilde{C}}^2 \tag{24}$$

where f varies between 0 and 1. The average value of the factor in square brackets is 2/3.

Combining Equations (21)–(24) produces an expression for the variance of the noise on a global reconstruction as a function of the variance of the noise on the input projection data.

$$\sigma_G^2 \cong \frac{\sigma_P^2}{3Vx^2} \tag{25}$$

The noise in local CT reconstructions is determined analogously. At the center of a centered cylinder, Equation (10) becomes

$$\Lambda\mu = S_n \sum_{r=1}^{V} \frac{\Delta\phi K_n(0)}{D} \tag{26}$$

$$K_n(0) = \sum_{j=-n}^{n} P(j\Delta\alpha) \cos (j\Delta\alpha) F_{n,j} \tag{27}$$

where V is the number of discrete source positions, $\Delta\phi$ is the angular rotation increment of the X-ray source ($\Delta\phi = 2\pi/V$), $K_n(0)$ is the convolved signal, D is the source-to-object radius, $P(j\Delta\alpha)$ is the project data, α is the X-ray fan angle, $\Delta\alpha$ is the X-ray angle increment, $F_{n,j}$ is the Smith [3] local convolution filter, and n is an integer that determines the number of samples in the PSF.

$$F_{n,j} = \frac{1}{\pi^2 x} \left(1 - \left(\frac{j}{n}\right)^2\right)^{m-1} \left(1 - (2m + 1) \left(\frac{j}{n}\right)^2\right) \tag{28}$$

Applying the rules of quadrature and multiplying by 2/3 to account for the interpolation smoothing produces the variance of $\Lambda[\mu]$.

$$\sigma_{L,n}^2 = \frac{\sigma_P^2}{3Vx^2} \frac{8S_n^2}{\pi^2 D^2} \sum_{j=-n}^{n} \left(1 - \left(\frac{j}{n}\right)^2\right)^{2m-2} \left(1 - (2m + 1) \left(\frac{j}{n}\right)^2\right)^2 \tag{29}$$

NONDESTRUCTIVE EVALUATION

TABLE 9.2. Comparative Noise between Global and
Local Filters for Equivalent Contrast Scaling.

Local CT Filter	n	$\sigma_{L,n}$
Five-point	2	6.07 σ_G
Nine-point	4	2.22 σ_G
Thirteen-point	6	1.23 σ_G

Using the values from Table 9.1 for the constant S_n, Equation (29) can be evaluated for the standard deviation of the noise in a local CT reconstruction expressed as a multiple of the noise in a comparably scaled global reconstruction. Table 9.2 lists noise for local reconstructions using filters with $n = 2, 4,$ and 6.

Noise on a local reconstruction made with a five-point filter is prohibitively high; noise on nine-point and thirteen-point reconstructions is acceptable. However, these reconstructions will be at a disadvantage compared to global reconstructions for features at the threshold of detectability. The threshold of detectability refers to the smallest object that can be detected at a given contrast level. It can be demonstrated that the threshold of detectability is directly proportional to the ratio of contrast to noise in CT [8].

As shown in Figure 9.3b, scaling does not improve threshold detectability because both contrast and noise are scaled. However, for objects having contrast well above the threshold level, the contrast-to-noise ratio is less important than the contrast perceived on the display.

RESOLUTION IN LOCAL CT

Resolution in CT images is a function of the PSF of the convolution function. However, the dependence of resolution on the PSF in local images is somewhat more complex than in global images, as illustrated in Figure 9.6. This plot is a superposition of the horizontal profile in Figure 9.5a passing through the five small, high-contrast cylinders embedded in the larger cylinders (dashed line) and the same profile in Figure 9.5b (solid line). As expected, features in the local reconstruction that are considerably wider than the filter PSF exhibit the characteristic positive/negative overshoot. However, as features become comparable in size to the PSF, the positive portions of the edge overshoots overlap, with the result that the contrast of the feature is increased, but the overall width is not. This produces a bonus in contrast detectability for small features.

Local CT images are ideally suited for dimensional analysis. The edges of objects that are larger than the PSF can be located accurately by cross-correlation with a mask that duplicates the overshoot. The dimensions of smaller features can be assessed by deconvolution with the PSF. Also, because the background is comparatively flat, many features can be identified with a simple threshold.

REGION OF INTEREST LOCAL CT

There are many reasons why it is sometimes preferable to scan and reconstruct only a region of interest (ROI) in a large object. Global CT requires data that encompass an object, including the unattenuated signal in the air outside it. Often, however, the object may be too wide to fit within the X-ray beam. In addition, the data load for a high-resolution scan of a large object may be prohibitive, or the object may be so dense that the X-ray intensity needed to penetrate it will overrange the X-ray detector in the air paths outside the object. If ROI data are reconstructed globally, severe shading is superimposed on the details of the image. Methods have been

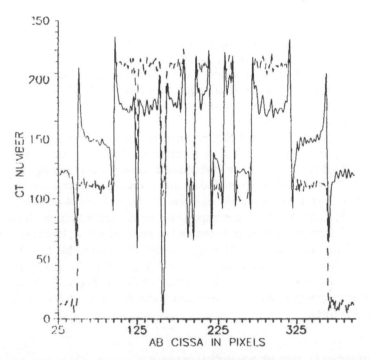

Figure 9.6 Superposition of one-dimensional horizontal profiles through reconstructed images of five high-contrast cylinders inside larger cylinders shown in Figures 9.5a and 9.5b: – – – global reconstruction; – – – – – local reconstruction, nine-point filter.

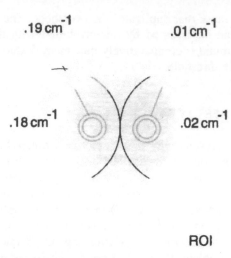

.19 cm^{-1} .01 cm^{-1}

.18 cm^{-1} .02 cm^{-1}

ROI

Figure 9.7 Cross section of mathematical test phantom containing spheres of different X-ray attenuation coefficients in which are embedded spherical shells; shaded circle represents ROI for which a CT data set is acquired.

devised to improve such images, but they usually involve additional time, data, and processing capacity, or a loss of resolution. Local CT reconstructions of ROI data are nearly indistinguishable from local reconstructions of complete data. Also, because local ROI images have good contrast and a flat background, they are ideal for automated detection and measurement procedures.

The advantages of local CT ROI reconstruction are illustrated in Figures 9.7–9.9. Figure 9.7 shows an ROI in a simulated phantom consisting of two large solid spheres in which are embedded spherical shells having contrast of 0.01 cm^{-1}. The ROI over which data are collected is represented by the dark circle. Figure 9.8a is the global reconstruction of one plane of this object, and Figure 9.8b is the corresponding local reconstruction. Shading and loss of contrast are evident in the global reconstruction; however, both shells and the boundaries of the large spheres are discernible in the local reconstruction. These differences are quantified by the plot in Figure 9.9, which is a superposition of the horizontal profile in Figure 9.8a passing through the spherical shells (dashed line) and the same profile in Figure 9.8b (solid line).

EXPERIMENTAL ROI RECONSTRUCTIONS

The following examples illustrate the performance of local ROI reconstructions on data obtained from the experimental system [9] shown

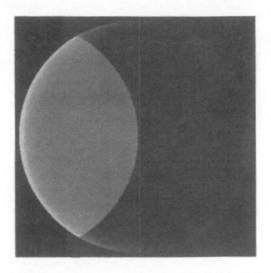

Figure 9.8a Global CT reconstruction of one plane thorugh test phantom shown in Figure 9.7.

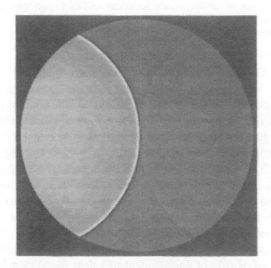

Figure 9.8b Local, nine-point filter, CT reconstruction of one plane through test phantom shown in Figure 9.7.

159

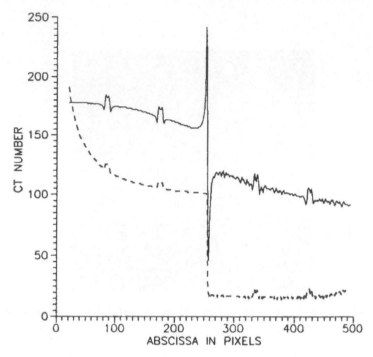

Figure 9.9 Superposition of one-dimensional horizontal profiles through spherical shells in the reconstructed images of Figures 9.8a and 9.8b; – – – global reconstruction; – – – – – local reconstruction, nine-point filter.

in Figure 9.1. Global reconstructions were made with the Shepp–Logan filter, and local reconstructions were made with the nine-point Smith filter. Figures 9.10 and 9.11 illustrate the extent to which shading in a global reconstruction can obscure a small feature near the circumference of an ROI image. Figure 9.10 is a sketch of a high-resolution Si_3N_4 phantom: the black circles represent holes drilled in the cylinder, and the dark circle marks the ROI. Figure 9.11a is a global reconstruction of this ROI. The shading in this image is so severe that the 25-μm hole is not discernible. Figure 9.11b shows the local reconstruction of the same phantom: the background is very flat, and the 25-μm hole is visible. Contrast of the other holes is also noticeably enhanced.

Figures 9.12 and 9.13 illustrate the extent to which shading can degrade the reconstructed density of uniform features near the circumference of an ROI image. Figure 9.12 is a sketch of a density phantom. It consists of Si_3N_4 GN-10 injection molding mix that contains 15.5 percent organic binder by weight. Embedded in the cylinder are five plugs of Si_3N_4 GN-10 injection molding mix containing percentages of organic binder ranging

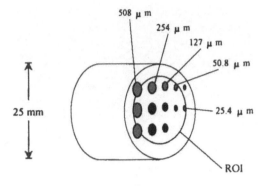

508 μm

254 μm

127 μm

50.8 μm

25 mm

25.4 μm

ROI

Si $_3$N$_4$ RESOLUTION
PHANTOM

Figure 9.10 Illustration of Si$_3$N$_4$ resolution phantom; shaded circle represents ROI for which CT data set is acquired.

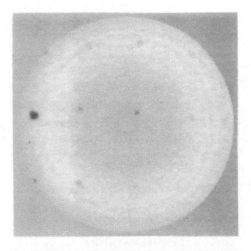

Figure 9.11a Global CT reconstruction of one plane through Si$_3$N$_4$ resolution phantom shown in Figure 9.10.

161

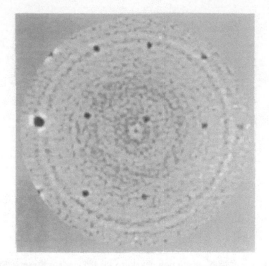

Figure 9.11b Local, nine-point filter, CT reconstruction of one plane through Si₃N₄ resolution phantom shown in Figure 9.10.

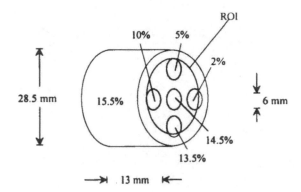

Si $_3$N $_4$
DENSITY PHANTOM

Figure 9.12 Illustration of Si₃N₄ density phantom; shaded circle represents for which CT data set is acquired.

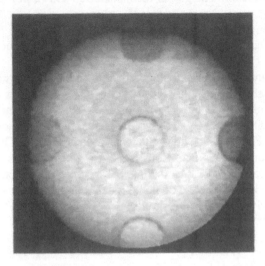

Figure 9.13a A global CT reconstruction of one plane through the Si_3N_4 density phantom shown in Figure 9.12.

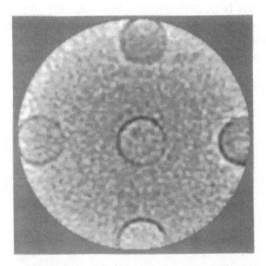

Figure 9.13b Local, nine-point filter, CT reconstruction of one plane through Si_3N_4 density phantom shown in Figure 9.12.

163

from 2 to 14.5 percent. The global reconstruction shown in Figure 9.13a is severely shaded, with the result that the plugs seem to have nonuniform density. The same slice of the local reconstruction shown in Figure 9.13b clearly shows that each plug has uniform density and that each has a different density.

Finally, Figures 9.14 to 9.16 illustrate the suitability of local CT for dimensional analysis from ROI reconstructions. Figure 9.14 is a simplified sketch of the cross section of an Si_3N_4 turbine rotor. Again, the dark circle represents an ROI. Figure 9.15a shows the global reconstruction of the turbine rotor, and Figure 9.15b shows the local reconstruction. The advantage of the local reconstruction is obvious in Figure 9.16, which is a superposition of a vertical column intersecting the rotor body taken from each reconstruction. The rotor edge could be located in the local reconstruction (dashed line) with a simple threshold.

SUMMARY

A comparison has been made between the relative merits of global and local computerized tomography (CT) for the applications of contrast discrimination, dimensional measurement, and region of interest (ROI) reconstruction. It has been demonstrated that local CT has advantages for contrast discrimination if (a) a wide range of densities is present in an image (b) there are features comparable in size to the point-spread-function of the system, and (c) a complete data set is available for only a smaller region of a large object. Although the standard deviation of noise reconstructed in local CT images has been found to be slightly higher than noise in com-

ROI

Figure 9.14 Sketch of simplified Si_3N_4 turbine rotor; shaded circle represents ROI for which CT data set is acquired.

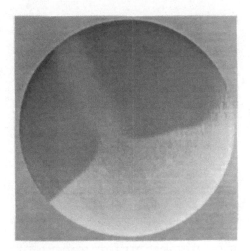

Figure 9.15a Global CT reconstruction of one plane through Si_3N_4 turbine rotor shown in Figure 9.14.

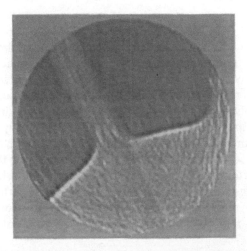

Figure 9.15b Local, nine-point filter, CT reconstruction of one plane through Si_3N_4 turbine rotor shown in Figure 9.14.

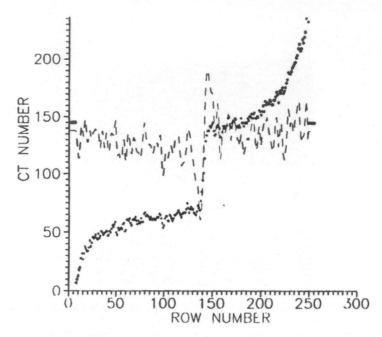

Figure 9.16 Superposition of one-dimensional vertical profiles through body of turbine rotor in reconstructed images of Figures 9.15a and 9.15b: **** global reconstruction; − − − local reconstruction, 9-point filter.

parably scaled global CT images, this does not always degrade detectability. For features with contrast well above the threshold of detectability, the relative increase in display contrast available in local images is advantageous. Because local CT reconstructs an edge-enhanced version of a global image upon a relatively flat background, automated detection and location of objects is fairly straightforward. This is especially true in ROI images. Global CT ROI images suffer severe shading, which greatly decreases the visibility of features and makes them difficult to isolate. We do not suggest that local CT is a substitute for global CT, especially for the task of absolute density determination; however, for the specified tasks, local CT is an attractive alternative. In addition, it is helpful to perform local CT reconstructions in conjunction with global reconstructions to identify quickly those in regions of an image likely to have interesting features.

ACKNOWLEDGEMENT

This chapter has resulted from work sponsored by USDOE/Fossil Energy/Advanced Research and Technology Development/Materials Program, under Contract W-31-109-ENG-38.

REFERENCES

1 Vainberg, E. I., Kazak, I. A., and Faingoiz, M. L. *Soviet J. Nondestructive Testing,* Vol. 17, 1981, p. 415.

2 Smith, K. T. and Keinert, F. "Mathematical Foundations of Computed Tomography," *Applied Optics,* Vol. 24, No. 3, 1985, pp. 3950–3957.

3 Faridani, A., Keinert, F., Natterer, F., Ritman, E. L., and Smith, K. T. "Local and Global Tomography," *Signal Processing, Part II: Control Theory and Applications,* F. A. Grunbaum, ed., Springer Verlag, New York, 1990, pp. 241–255.

4 Faridani, A., Ritman, E., and Smith, K. "Local Tomography," *SIAM J. Appl. Math,* Vol. 52, No. 2, 1992, pp. 459–484.

5 Feldkamp, L., Davis, L., and Kress, J. "Practical Cone-Beam Algorithm," *J. Opt. Soc. AM, A,* Vol. 1, No. 6, 1984, pp. 612–619.

6 Ramachandran, G. N. and Lakshminarayanan, A. V. "Three-Dimensional Reconstruction from Radiographs and Electron Micrographs," *Proc. National Academy of Science,* Vol. 68, 1971, pp. 2236–2240.

7 Shepp, L. A. and Logan, B. F. "The Fourier Reconstruction of a Head Section," *IEEE Trans. Nuc. Sci.,* NS21, 1974, pp. 21–43.

8 Sivers, E. A. and Silver, M. D. "Performance of X-ray Computed Tomographic Imaging Systems," *Materials Evaluation,* Vol. 48, 1990, pp. 706–713.

9 Ellingson, W. A., Vannier, M. W., and Stintor, D. P. "Application of X-ray Computed Tomography to Ceramic/Ceramic Composites," *Characterization of Advanced Materials,* W. Altergath and E. Henneke, ed., Plenum Press, New York, 1991, pp. 9–25.

CHAPTER 10

Nondestructive Evaluation of Ceramic and Metal Matrix Composites for NASA's Materials Programs

EDWARD R. GENERAZIO[1]

OVERVIEW

IN a preliminary study, ultrasonic, X-ray opaque and fluo-rescent dye penetrant nondestructive techniques were used to evaluate and characterize ceramic and metal matrix composites. Techniques are highlighted for identifying porosity, fiber alignment, fiber uniformity, matrix cracks, fiber fractures, unbonds or disbonds between laminae, and fiber-to-matrix bond variations. The nondestructive evaluations (NDE) were performed during processing, after processing, and after thermomechanical testing. Specific examples are given for Si_3N_4/SiC (SCS-6 fiber), $FeCrAlY/Al_2O_3$ (sapphire fiber), Ti-15-3/SiC (SCS-6 fiber) materials, and Si_3N_4/SiC (SCS-6 fiber) actively cooled panel components. Results of this study indicate that the choice of the NDE tools to be used can be optimized to yield a faithful and accurate evaluation of advanced composites.

INTRODUCTION

The next generation of commercial aircraft will incorporate advanced aerodynamic and propulsion concepts that will require materials systems to be robust at extremely high temperatures and operating loads. NASA's advanced materials program focuses on the development of these enabling materials and structures for rotorcraft, ultrahigh bypass ratio engines for subsonic aircraft and supersonic high-speed civil transport (HSCT). The

---

[1]NASA Lewis Research Center, Cleveland, OH, U.S.A.; current address NASA Langley Research Center, Hampton, VA, U.S.A.

development of a demonstration of critical components for the HSCT is being pursued. Ceramic, intermetallic, and metallic matrix composites (CMCs, IMCs, and MMCs) are being screened and evaluated [1–4] for their use as base materials for critical components. Nondestructive evaluation (NDE) plays an important dual role in the development of these materials and in the evaluation and certification of components before and during use. It is a challenge for the nondestructive evaluation community to establish inspection (NDI) and evaluation (NDE) techniques that are accepted, reliable, and standardized for these enabling advanced composite systems.

Materials development and tailoring can be greatly accelerated by placing NDE at critical processing stages where it can play an active and important role in the development process. NDE can assist in material properties screening, in materials selection, and for determining the overall quality of finished components. NDE may occur in conjunction with or separately from proof testing. The component integrity after shipping, installation, and repair will also need to be addressed by NDE. Equally important is the consideration and inclusion of nondestructive inspection (NDI) requirements at the component design stage. The components must be designed so that a nondestructive inspection can be performed. Including NDI requirements in the component design will avoid many of the inspection complications, problems, and uncertainties encountered in other aerospace systems. This level of NDI may also include nonintrusive sensors that will monitor key variables that indicate the health of the component during use. Fiber and matrix cracks, delaminations, and fiber-to-matrix interface variations are material features that affect the composite's strength and toughness and, therefore, are prime candidates for quantitative nondestructive evaluation.

There are many NDE tools that can be used for evaluating materials. The choice of the tool to use for the inspection is one of the most important decisions during the inspection procedure. For example, an eddy current instrument will not be sensitive to the presence of cracks in an electrically nonconducting material. Although this is an extreme and obvious example, there are numerous instrument-material mismatches that are more subtle and, as such, must be avoided. It is appropriate to establish which methods, instruments, and techniques are best suited for the NDI and NDE at each stage of the combustor liner and nozzle development, certification, installation, repair, and use.

This work is a preliminary study that highlights standard NDE techniques for advanced composite systems. Five accepted NDE techniques were used; ultrasonic "C" and surface wave scans, conventional and microfocus X-ray film radiography, and fluorescent and X-ray opaque penetrants [5,6]. These techniques can be used to characterize fiber fractures, matrix cracks, porosity variations, delaminations, unbonds, debonds, and fiber-to-matrix interface variations. A variation on the data acquisition configuration and data analysis for a conventional c-scan system yields images of subsurface fibers and their "degree of bonding" to the matrix. This data acquisition and data analysis process extracts the subsurface reflection coefficient at the fiber-to-matrix interface from the surface reflection coefficient obtained from a standard c-scan. This deep reflection coefficient imaging (DERCI) [7] technique is in a development stage and is also shown here with its very promising preliminary results. Eddy current [5] and other more advanced techniques, such as thermal diffusivity imaging [8,9], shearography [10], speckle interferometry, scanning acoustic microscopy (SAM) [11] scanning electron acoustic microscopy (SEAM) [12,13], and computer-aided X-ray tomography (CAT) [14], will be addressed in a separate article to be published at a later date.

The chapter is organized into two sections: 1) NDE for processing and quality, and 2) NDE for degradation. The goal of this work is to show how NDE tools can be used effectively to impact the processing procedures, evaluate the finished component or test piece, and evaluate the degree of degradation that occurs from use or thermomechanical testing.

NDE FOR PROCESSING AND QUALITY

CERAMIC MATRIX COMPOSITES

We begin our study with the development of an Si_3N_4/SCS-6 [15] panel that contains coolant channels (Figure 10.1). The actively cooled panel is made up of a monolithic, 10.0×10.0 cm Si_3N_4 slab that is laminated to a cross-ply composite panel Si_3N_4/SCS-6 [0–90]. The laminated panel is 0.335 cm thick. The monolithic panel has 0.05-cm wide and 0.05-cm high coolant channels. The channel walls are 0.05 cm thick. During the development of this panel, the interlaminar bonding procedure is evaluated by bonding two similar but monolithic plates. The bond between planar layers may be evaluated with radiographic and through transmission, focused 10 MHz, ultrasonic c-scan analysis; the results are shown in Figures 10.2 and 10.3, respectively. A low X-ray intensity, i.e., high X-ray attenuation, corresponds to the dark region in the X-ray film positive. The dark vertical bands in the radiograph correspond to the walls of the coolant channels. The channels are uniform in shape, and there are no chips, cracks, or occlusions present. There are two lighter vertical bands at areas where poor reaction of the bonding material produced low densities. (A subsequent modification of the manufacturing process eliminated these poorly reacted areas.) The ultrasonic c-scan shows these poorly reacted regions more

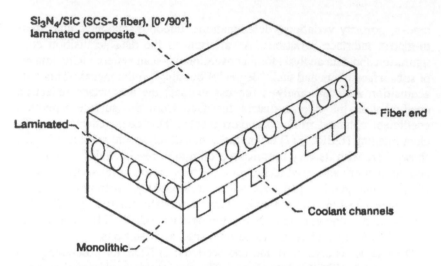

Figure 10.1 Schematic diagram showing structure of composite, actively cooled panel.

Figure 10.2 Radiograph of monolithic Si_3N_4, actively cooled panel.

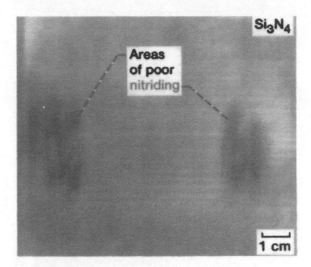

Figure 10.3 Ultrasonic, 10 MHz, through transmission c-span of monolithic Si₃N₄, actively cooled panel.

clearly and to be somewhat larger than that observed in the radiograph. The condition of the channels cannot be determined from the ultrasonic information.

Figures 10.4 and 10.5 show the radiographic and focused 10 MHz, ultrasonic c-scan results for a Si_3N_4/SCS-6 [0°] panel. Severe microcracking is observed in the radiograph as short light horizontal lines. The microcracks are most visible at the walls of the coolant channels. Note that, in the enlarged region of the radiograph, the crack lengths are not limited to the channel wall thickness but actually are several channels wide. The short dark vertical lines scattered throughout the image each have a short horizontal dark line at their central regions. These short dark vertical regions are regions where the channel has decreased in width. The dark horizontal line at the center of each of these regions, when examined at high magnifications, has been found to be an optical illusion that occurs where the channel walls have the closest approach to each other. At the region of closest approach, some channels are partially blocked by this narrowing of the channel. Several channels had adjacent walls that were of sufficient thickness that the walls were touching and apparently completely blocking the channel. The ultrasonic image reveals a considerable amount of disordered mottling in the structure. The radiograph does exhibit density variations. There is no correlation between the mottling structure in the ultrasonic c-scan image and the crack structure observed in the radiograph. This mottling is due to porosity and bond variations between the laminated layers [16–22].

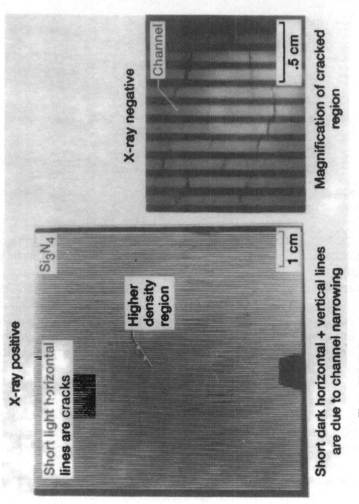

X-ray negative

Channel

.5 cm

Magnification of cracked region

X-ray positive

Si₃N₄

Short light horizontal lines are cracks

Higher density region

1 cm

Short dark horizontal + vertical lines are due to channel narrowing

Figure 10.4 Radiographs of Si₃N₄/SCS-6 [0 degree] actively cooled panel.

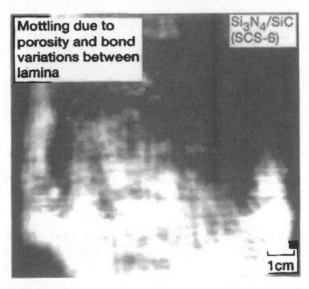

Figure 10.5 Ultrasonic, 10 MHz through transmission c-scan of Si₃N₄/SiC SCS-6 [0 degree] actively coolant panel.

In the previous example, the fiber orientation is not easily determined. Figures 10.6 and 10.7 show the radiographic and focused 10 MHz, ultrasonic, c-scan images of an Si_3N_4/SCS-6 [0/90°] composite panel that does not have coolant channels. The radiograph reveals a bowed fiber system. These bowed patterns are not actual individual bowed fibers but are X-ray "shadows" of a collection of fibers that are bowed. The shadows forming the bowed pattern are due to fiber volume density variations through the thickness and perpendicular to the X-ray film plane. This is manifestation of the Moiré effect [18,23] found in composite systems. The density of the

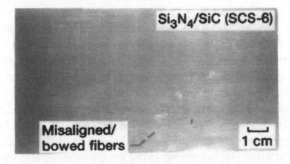

Figure 10.6 Radiograph of Si₃N₄/SiC SCS-6 [0–90 degree] panel.

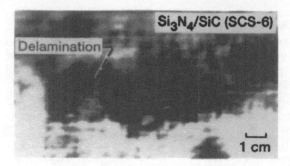

Figure 10.7 Ultrasonic, 10 MHz, through transmission c-scan of Si₃N₄/SiC SCS-6 [0–90 degree] panel.

system is uniform, as indicated by the uniform gray level or shading in the radiograph. The dark areas in the ultrasonic image also yielded reflected ultrasonic signals that corresponded to reflections from the laminated boundary; these dark areas are regions having poor or no bond between laminated layers.

Fibers need only be well ordered and spaced uniformly in the plane of an Si_3N_4/SCS-6 [0/90] plate in order to yield relatively uniform radiographic and ultrasonic images (Figure 10.8). However, if there are any systematic variations, such as ordered through-the-thickness buckling (due to processing) of the fibers (Figure 10.9), then the ultrasonic image will reveal these systematic fiber spacing variations. The fiber buckling can be readily observed in the photomicrograph (Figure 10.9) by comparing the spacing between the bottom two rows of fibers that are perpendicular to the cross section. Note that this spacing varies dramatically from the center to the right edge of the sample. In this case, the fibers are forming a crude diffractive/refractive acoustic lens (Figure 10.9) that is observed as a bright spot in the ultrasonic image. No indication of this fiber buckling is seen in the radiograph.

An SiC/SiC (Nicalon fiber) laminated, two-dimensional, woven composite yields very complicated radiographic and focused 10-MHz, ultrasonic c-scan (Figure 10.10) images. The radiograph reveals a cross hatch pattern that varies in density. This variation may be due to density variations in the matrix material or registration variations between the woven layers. The dark circular regions that are uniformly spaced are due to local density variations that have been produced during processing. These circular density variations are also visible in the ultrasonic image. The distorted or blurry appearance of the ultrasonic images is due to refractive and diffractive scattering [17] occurring at the rough surfaces and woven fibers. There is little additional information obtained from the 10-MHz, ultrasonic c-scan.

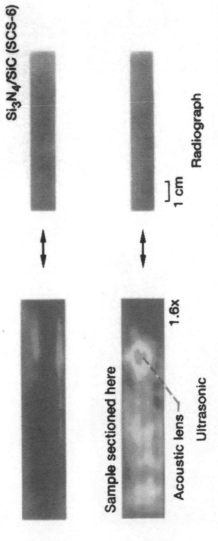

Figure 10.8 Ultrasonic, 10 MHz, through transmission c-scan and radiographs of similarly produced Si₃N₄/SiC SCS-6 [0–90 degree], test bars.

177

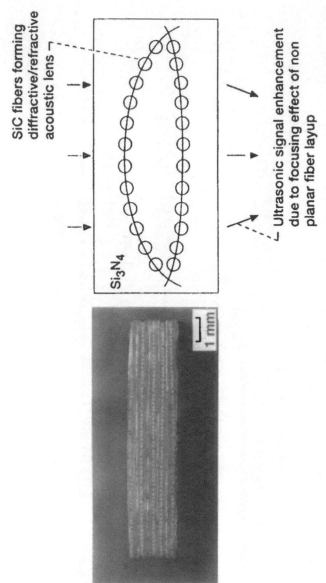

SiC fibers forming diffractive/refractive acoustic lens

Si₃N₄

Ultrasonic signal enhancement due to focusing effect of non planar fiber layup

1 mm

Figure 10.9 Cross section of Si₃N₄/SiC SCS-6 [0–90 degree] test bar at region showing large ultrasonic transmission.

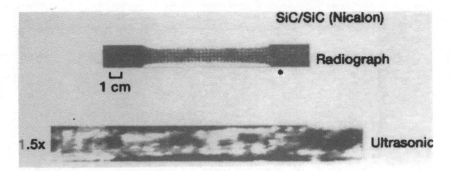

Figure 10.10 Ultrasonic, 10-MHz through transmission c-span and radiograph of SiC/SiC (Nicalon) woven laminated composite.

METALLIC MATRIX COMPOSITES

During the development of potential metal matrices, the processing procedures are optimized by a combination of methods. One technique developed for consolidating metallic glass powders requires the use of a molybdenum or niobium encapsulating can for hot isostatic pressing (HIP). Radiographic (Figure 10.11) and ultrasonic (Figure 10.12) imaging can be performed while the consolidated material, NbYSi, is still capsulated. The radiograph reveals an extensive amount of cracking throughout the consolidated material. Regions of increased porosity can be readily identified as light cloud-like areas in the radiograph and dark areas in the ultrasonic image. The crack density increases with an increase in porosity.

The degree of consolidation was monitored in Ti-15-3/SiC (SCS-6) fiber composites with radiographic (Figure 10.13) and ultrasonic imaging (Figure 10.14). This panel was HIP-ed in two steps. The second processing

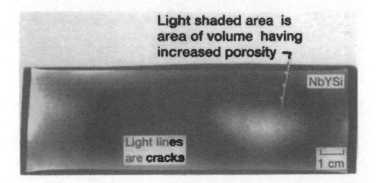

Figure 10.11 Radiograph of encapsulated NbYSi matrix material.

Figure 10.12 Ultrasonic, 10-MHz, through transmission c-scan of encapsulated NbYSi matrix material.

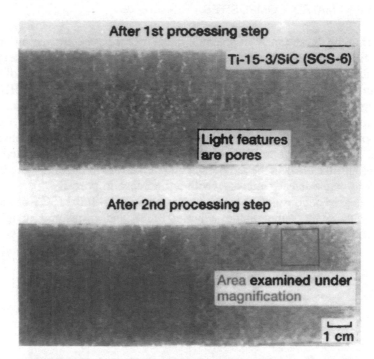

Figure 10.13 Radiographs of Ti-15-3/SiC (SCS-6) composites after first and second processing steps.

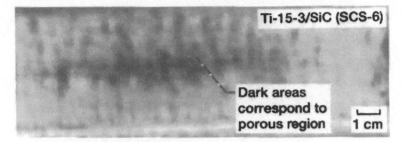

Figure 10.14 Ultrasonic, 10-MHz, through transmission c-scan of Ti-15-3/SiC (SCS-6) composite after the first processing step.

step has removed much of the porosity (Figure 10.13). A close examination of the radiograph reveals fine fiber fractures (Figure 10.15). The intensity of the X rays at the fracture sites is low. This low intensity or high X-ray attenuation at the fracture sites is an indication that the matrix material has infiltrated into these fracture sites, as shown in Figure 10.15. An increase in the X-ray intensity at the fracture sites would indicate that these fracture sites contained voids. The fractures occurred during the ductile part of the HIP stage and are not due to the mismatch between the coefficients of thermal expansion of the fiber and matrix materials. The initial infiltration of the matrix into the fiber fracture sites was radiographically observed after the first HIP stage.

Composites with three phases yield complex radiographic and ultrasonic data. A Ti-15-3 plate containing unidirectional SiC (SCS-6) fibers and a molybdenum wire mat represents a simple three-phase composite system. Groups of molybdenum wires can be identified in the focused 10-MHz, ultrasonic c-scan image (Figure 10.16) as faint uniformly spaced vertical lines. Here, each group of mat wires act as an ultrasonic line scatter. Dark horizontal bands in the upper and lower parts of the ultrasonic image are due to the slight bunching of the horizontal SiC fibers, i.e., areas having high fiber volume density. The central region of the ultrasonic image also exhibits some horizontally oriented mottling. This mottling is due to similar fiber volume density variations, except that these fiber volume density variations are not as ordered as those near the upper and lower edges of the plate. The radiograph shows both the orientation and structure of the horizontal SiC fibers and the vertical molybdenum wires that make up the mat (Figure 10.17). Broken, twisted, and bent molybdenum wires are easily identifiable and occur throughout the system. Fiber-rich banded zones can also be identified, but not easily, in the radiograph.

Ordered or systematic changes in the fiber volume density or orientation can produce artifacts or Moire [17,23] patterns in the radiographs. For example, if in one layer, the fiber spacing, S1, is slightly different from the

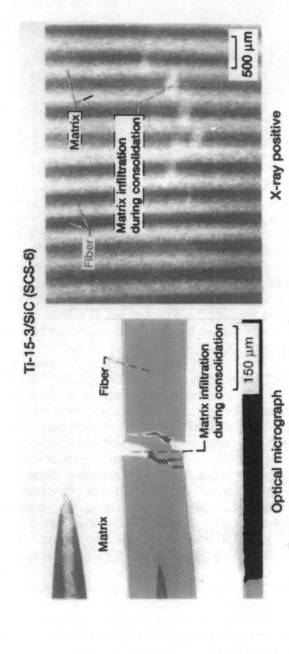

Figure 10.15 Metallograph and enlarged X ray positive of region highlighted in Figure 10.13.

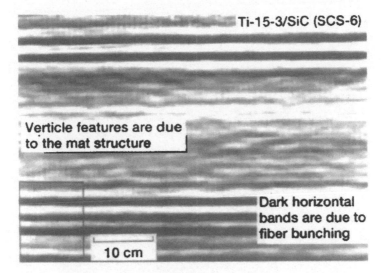

Figure 10.16 Ultrasonic, 10-MHz, through transmission c-scan of Ti-15-3/SiC (SCS-6) composite.

fiber spacing, S2, in another layer, then a Moiré pattern will be observed in the radiograph. The radiograph of a copper matrix system containing tungsten wires (Figure 10.18) reveals broad vertical dark Moiré bands. (This is not an advanced high-temperature composite material, but it does dramatically show an artifact that occurs in all continuous fiber composites to some degree.) In addition, if the fiber layers are also off-axis from each other, then the Moiré bands will also be off-axis.

NDE FOR DEGRADATION

CERAMIC MATRIX COMPOSITES

There is a wide range of degradation or failure mechanisms that can rapidly affect the properties of CMCs. The primary mechanisms that affect the properties of CMCs are erosion, interfacial oxidation, and also impact, thermal, mechanical, and acoustical stresses that initiate fiber, fiber coating, fiber-to-matrix interface, and matrix fractures. The toughness of these brittle matrix composites is strongly dependent on the character and strength of the fiber-to-matrix interface. All of the degradation mechanisms mentioned above can lead to a degradation of the interfacial shear strength. Additionally, interfacial property changes due to thermal, mechanical, or chemical treatments are quite different. Thermally driven

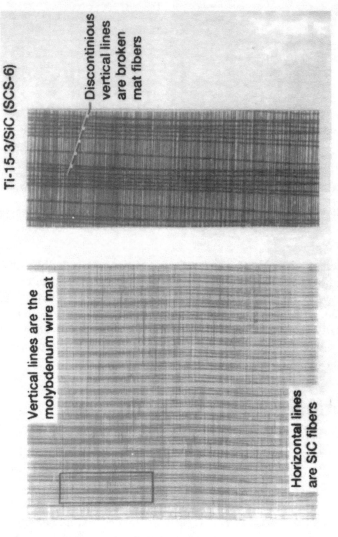

Ti-15-3/SiC (SCS-6)

Discontinious vertical lines are broken mat fibers

Vertical lines are the molybdenum wire mat

Horizontal lines are SiC fibers

Figure 10.17 Radiograph of highlighted region in Figure 10.16.

184

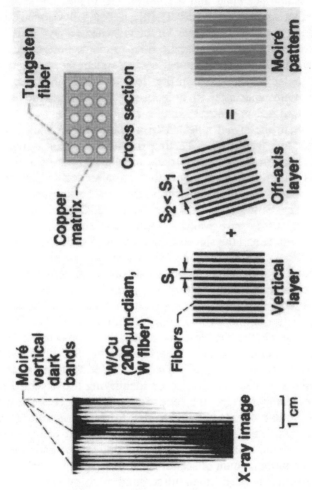

Figure 10.18 Radiograph and illustrations showing Moiré effect in W/Cu composite.

185

interfacial changes result in drastic changes, via loss or replacement of original interface material [24,25]. In contrast, mechanically driven interface changes result in debonding, chipping, and movement or redistribution of the interface material [26]. The following work in this section focuses on the characterization of this interface.

A new ultrasonic scanning and evaluation technique, known as deep reflection coefficient imaging (DERCI) [6], is being developed to characterize the fiber-to-matrix interface. An image generated by DERCI characterizes the reflection coefficient at the fiber-to-matrix interface. DERCI may be implemented on a conventional c-scan system by time gating and acquiring the reflected signals from the fiber-to-matrix interface. These reflected signals are superimposed, at focused 50 MHz, onto the reflection from the front surface of the test sample to form a signal with multiple peaks. Most often, the signal reflected from the front surface is not collected, but used as a trigger for collecting signals that occur at much later times, e.g., echoes from the back surface or interlaminar delaminations. However, it is this front surface signal that contains detailed information about the bonding of the fibers near the surface of the sample. There is a distinct difference between surface wave and DERCI evaluations. At a fixed frequency, surface wave evaluation interrogates the surface and near surface of the sample, and the signals observed are dependent on the depth of the scatter (fiber). In contrast, the DERCI image can interrogate deep into the structure of the composite ply. In its optimum configuration, the DERCI image is independent (except for geometric diffraction corrections) of the distance between the sample and ultrasonic transducer.

A brief description of the DERCI interpretation follows. The ultrasonic reflection coefficient at the fiber-to-matrix interface is expected to decrease with an increase in bonded contact area between the SiC fiber and Si_3N_4 matrix. The two extreme cases of perfectly bonded and completely unbonded SiC fibers provide the insight for identifying the expected reflection coefficient changes. Here, the acoustic impedances between the SiC and Si_3N_4 are similar when compared to that for a void and Si_3N_4. If the SiC fibers are completely unbonded, then the ultrasound will see the SiC fiber region as a void (maximum impedance difference) and, therefore, will yield a maximum amount of scattering at this type of interface. If the fibers are perfectly bonded, then ultrasound with wavelengths much greater than the interface thickness will pass across this interface into and through the fiber with little scattering at the interface. In terms of imaging, a well-bonded fiber with its low reflection coefficient at the interface will be difficult to observe acoustically, while a poorly bonded fiber with its high reflection coefficient will be easily observed acoustically.

The interface reflection coefficient will vary uniformly with changes in the interface between the fiber and the matrix. There are many interfacial characteristics, such as thickness, density, contact area between the fiber and matrix, and modulus, that will affect the acoustic impedance of this interface [6]. Here, we will focus on mechanical decoupling and thermally driven oxidation of the bonded contact area of the interface and its interrelationship with the interfacial shear strength.

An Si_3N_4/SiC [0] (SCS-6 fiber) composite was tested in tension until matrix cracking was visually observed. The specimen was then unloaded, and a NDE was performed. A region of the specimen was identified that did not visually exhibit matrix cracking. A radiograph of this area reveals a uniform structure (Figure 10.19). A microscopic evaluation of the radiograph does not reveal the presence of any microcracks. However, a 50 MHz, ultrasonic DERCI image (Figure 10.20) reveals a detailed subsurface structure. The individual subsurface fibers of the first ply can be identified as horizontal features across the image. The vertical and near vertical features are cracks in the matrix material. Note that, along and near the matrix cracks, there are areas in the image where the fibers appear to be clear or less blurry. The fibers in these areas have decoupled from the matrix material and, in doing so, have formed nearly perfect cylindrical ultrasonic scattering boundary. This rather clean boundary prohibits any secondary or blur-inducing scattering of the wave that would have occurred if the ultrasound travelled into the SiC fiber (Figure 10.21). A fluorescent dye penetrant makes these matrix cracks visible under an ultraviolet light (Figure 10.22).

A zinc iodide X-ray opaque penetrant was used to determine the extent of the debonding of the fibers from the matrix material. Immediately after the zinc iodide is applied to the sample, a radiograph is taken (Figure 10.23). The radiograph reveals numerous cracks as light near vertical lines. These cracks can also be identified in the DERCI image (Figure

Si_3N_4/SiC (SCS-6)

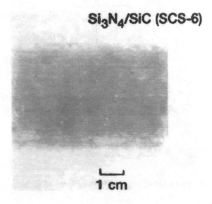

1 cm

Figure 10.19 Radiographed section of Si_3N_4/SiC SCS-6 [0] RBSN tensile test bar.

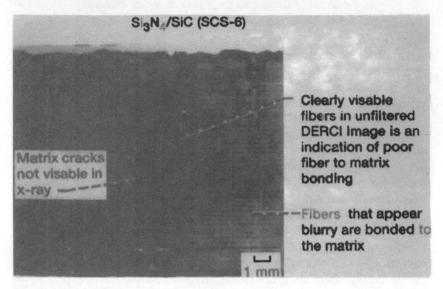

Figure 10.20 Ultrasonic DERCI unfiltered image at 50 MHz of section of Si₃N₄/SiC SCS-6 [0] RBSN tensile test bar.

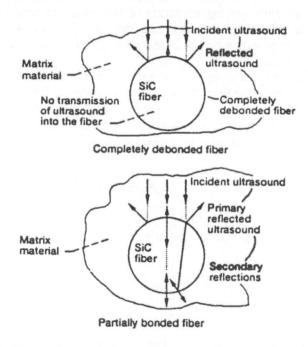

Figure 10.21 Schematic diagrams indicating possible source of secondary image blurring scattering.

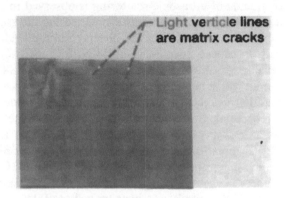

Figure 10.22 Fluorescent penetrant image of section of Si₃N₄/SiC SCS-6 [0] RBSN tensile test bar.

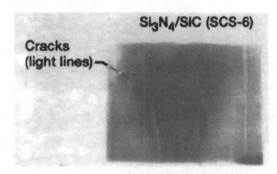

Figure 10.23 Radiograph of section of Si₃N₄/SiC SCS-6 [0], RBSN tensile test bar immediately after application of zinc iodide penetrant.

10.20). One hour after the initial application of penetrant, the sample is radiographed again (Figure 10.24) in an identical fashion. The sharp crack-like features have become faint and broad. This is an indication that the penetrant is slowly spreading throughout the sample. The wicking of the penetrant stopped at about one hour after its application.

An extensive amount of ultrasonic scattering is observed in a 10-MHz, ultrasonic c-scan (Figure 10.25). The dark areas in the ultrasonic image correspond to the areas where cracks are observed in the DERCI image.

Two identically produced specimens of Si_3N_4/SiC (SCS-6) [0] were ultrasonically evaluated using the DERCI technique. The DERCI images are spatially filtered with a two-dimensional high pass filter before further evaluation. This filter extracts the important spatial high-frequency components (i.e., reflections from the fiber-to-matrix interface) from the reflections from porosity, which are generally observed as mottling in ceramics [16–22] (Figures 10.5 and 10.7). The DERCI results reveal similar features in both samples before heat treatment (Figure 10.26). The vertical features in these images are due to reflections at the subsurface fiber-to-matrix interfaces. One sample was heat treated at 600°C for 100 h in flowing oxygen to degrade the fiber-to-matrix interface (via oxidation) [24]. This heat treatment decreased the interfacial shear strength from 18 ± 4 MPa to 0.8 ± 0.4 MPa [24]. After heat treatment, there is a marked change in the DERCI image. After heat treatment, the reflections from the interface are more intense so that the fibers become more visible. This change in the DERCI results before and after heat treatment can be quantified. The histogram maximum of the gray scale of each of the images may be used to quantify the DERCI results (Figure 10.27). A low value for the histogram maximum indicates that there is considerable amount of scattering at the fiber-to-matrix interfaces per unit area. A high maximum value indicates that the fibers are difficult to observe and that the image is rather bland or smooth. Ideally, if the fibers were perfectly

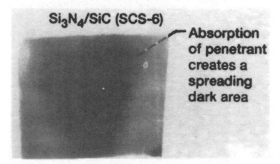

Figure 10.24 Radiograph of section of Si_3N_4/SiC SCS-6 [0] RBSN tensile test bar one hour after application of zinc iodide dye penetrant.

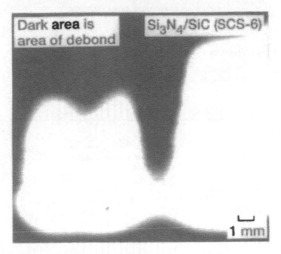

Figure 10.25 Ultrasonic, 10 MHz, through transmission c-scan of section of Si₃N₄/SiC SCS-6 [0] RBSN tensile test bar.

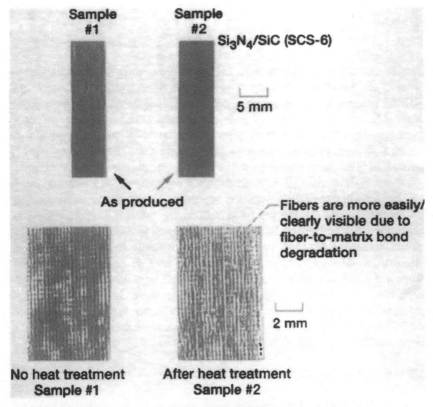

Figure 10.26 DERCI images of Si₃N₄/SiC SCS-6 [0] RBSN at 50 MHz before and after heat treatment.

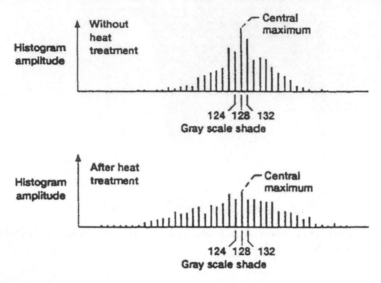

Figure 10.27 Histograms of DERCI images of Si$_3$N$_4$/SiC SCS-6 [0] RBSN samples before and after heat treatment.

bonded and, therefore, essentially unobservable (except for the small variation in the acoustic impedance between the fiber and matrix), then the DERCI images would contain predominantly one shade, 128, in the gray scale. The gray scale maximums for the untreated and heat treated specimens are 1.9 are 3.3 (arbitrary units), respectively. This corresponds to about a 75 percent change in histogram amplitudes. These values can be correlated with the respective interfacial shear strengths. If the interfacial shear strength is strongly dependent on the contact area between the fiber and matrix, then the DERCI results, which are also sensitive to changes in this contact area, are images of variations in interfacial shear strength. Since the fiber and matrix acoustic impedances are not identical, then the interrelationship between the DERCI results and the interfacial shear strength, which to first approximation is assumed to be a linear relationship, will contain a calibration or scale factor. The radiographs before and after the heat treatment (Figure 10.28) show essentially similar results.

These ultrasonic results are similar [7] to those found for a series of CMC samples (Figure 10.29) produced to have a wide range of interfacial bond strengths (Figure 10.30).

METALLIC MATRIX COMPOSITES

An FeCrAlY/Al$_2$O$_3$ [0] (single crystal sapphire fiber) composite bar was tested in tension at 300 K until failure. Fine cracks in the fibers are barely

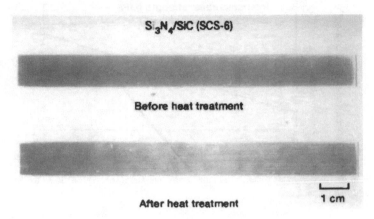

Figure 10.28 Radiographs of Si_3N_4/SiC SCS-6 [0] RBSN sample #2 before and after heat treatment.

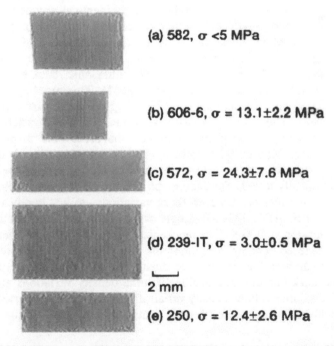

Figure 10.29 DERCI images of Si_3N_4/SiC SCS-6 [0] RBSN samples having a range of interfacial shear strengths.

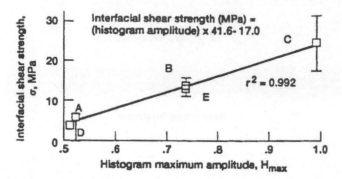

Figure 10.30 Histogram maximum amplitude versus interfacial shear strength for the Si₃N₄/SiC SCS-6 [0] RBSN samples.

visible under magnification of the radiograph shown in Figure 10.31. The fibers fractured far away from the test area where the sample ultimately fractured. The horizontal dark and light bands are due to changes in the fiber volume density through the thickness of the sample. This banding, even though it is not cyclic, is produced in an identical way as the Moiré patterns discussed previously. Similar bands have been found in untested samples. A 50-MHz, ultrasonic surface wave examination of the region having cracked fibers also reveals the extent and location of these cracked fibers (Figure 10.32). A focused 30-MHz, ultrasonic, back-echo c-scan reveals a mottled matrix material (Figure 10.32). This mottling is related to the original surface condition (i.e., before grinding the surface smooth) of the sample. The mottling is most likely due to porosity variations that are dependent on the manufacturing process that creates the "as-produced" rough surface. Figure 10.33 optically shows the as-produced and "as-ground" surface structures. All samples have as-ground surfaces before being mechanically tested. The radiograph (Figure 10.31) does exhibit some mottling in the matrix; however, these variations are just barely visible.

Another FeCrAlY/Al₂O₃ [0] (single crystal sapphire fiber) sample was tested in tension at 1100 K. The mechanical test was interrupted before the sample fractured. The light region in the X-ray (Figure 10.34) results correspond to an area that has thinned during the test. The same horizontal light and dark banding due to fiber volume density variations are identified. These fiber volume density variations are also easily observed in the focused 10-MHz, ultrasonic, through transmission c-scan (Figure 10.34).

DISCUSSION AND SUMMARY

Table 10.1 provides a summary of the NDE results. This table may be used as a guide to assist in the nondestructive evaluation of ceramic, in-

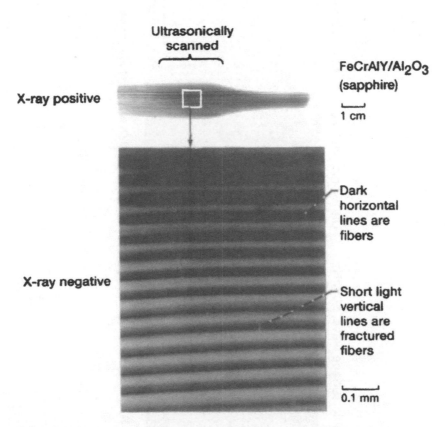

Figure 10.31 Radiograph and radiographic enlargement of FeCrAlY/Al203 (sapphire fiber) composite.

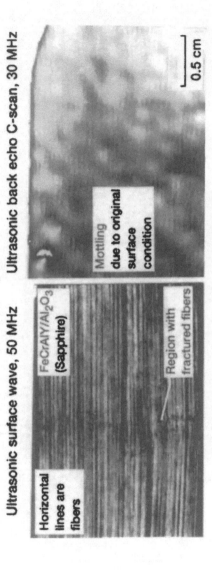

Figure 10.32 Ultrasonic surface wave and back-echo images of FeCrAlY/Al₂O₃ (sapphire fiber) composite.

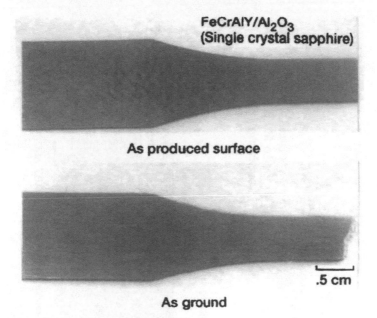

Figure 10.33 Optical images of "as-produced" and "as-ground" FeCrAlY/Al₂O₃ (sapphire fiber) composite.

termetallic, and metal matrix composites. Table 10.1 indicates that, by choosing the appropriate NDE technique, flaws such as fine fiber cracks, fiber-to-matrix interface variations, porosity, matrix cracks, interlaminar disbonds or unbonds, fiber misalignment, ply registration, and orientation can be easily identified with conventional NDE techniques and existing NDE systems. In addition, NDE can further assist and guide the development of composites by providing feedback information on subsurface material features that are present at various stages of processing and degradation. The ultrasonic DERCI technique has been identified as a technique for evaluating "degree of bonding" between the fiber and matrix in ceramic matrix composites. Artifacts observed in ultrasonic and radiographic images have been observed and attributed to Moiré and fiber buckling mechanisms.

ACKNOWLEDGEMENTS

The author would like to gratefully acknowledge all of the materials research scientists in the Materials Division, NASA Lewis Research Center that produced, provided, and processed the composite materials

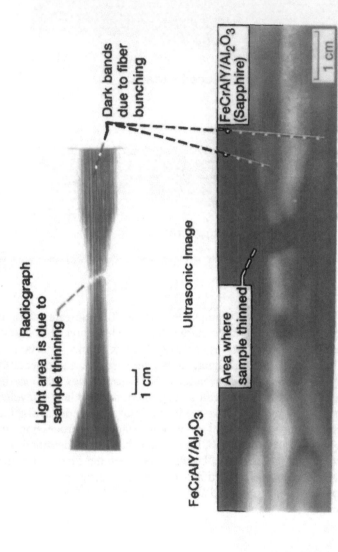

Figure 10.34 Ultrasonic, 10-MHz, through transmission c-scan and radiograph of FeCrAlY/ Al₂O₃ (sapphire fiber) composite tensile test bar.

TABLE 10.1.

	CMC	MMC
Matrix cracks	DERCI, FP	X, UC, FP, XP[a]
Fiber fractures	X[b]	US, X[b,c]
Delaminations	UC, XP	No samples with delaminations
Porosity	X, UC	X, UC
Fiber-to-matrix interface variations	DERCI, US	UC, APSS[d]
Fiber bunching, and layup registration	X	UC, X
Fiber alignment	X	X, UC
Fiber through-the-thickness buckling	UC	Not explicitly observed

APSS	Angular power spectrum scanning
DERCI	Deep reflection coefficient imaging
FP	Fluorescent penetrant
UC	Ultrasonic c-scan
US	Ultrasonic surface wave
X	X-ray (conventional/microfocus)

[a]Penetrants are used routinely for locating surface breaking cracks in metals [5].

[b]If the fracture planes are in the parallel direction of travel of the X-ray photons, then these may be observable when the fiber crack is widened [27].

[c]In a recent separate unpublished study, fractured fibers in FeCrAlY/Al_2O_3 (sapphire) composites were unobservable using conventional radiography. This is believed to be due to the complex fracture structure and X-ray absorption properties of sapphire.

[d]Ultrasonic through transmission and back-echo c-scans, and angular power spectrum scanning (APSS; [17]), have been used to identify stress relaxation and fiber ratcheting in heat treated, thermal cycled MMCs. However, the physical scattering mechanism responsible for the interrelationship is not well understood at this time.

used in this work. Without their interest and assistance, this work would not have been possible. I would like to thank Dr. Rebecca A. MacKay and John W. Pickens for the Ti-15-3/SiC materials; Susan L. Draper and Darrell J. Gaydosh for the FeCrAlY/Al₂O₃ materials; Dr. Ramakrishna Bhatt, Andy J. Eckel, Dr. Jeffrey I. Eldridge, and Dennis S. Fox for the RBSN/SiC and SiC/SiC composites; Marc R. Freedman for the Si₃N₄/SiC actively cooled panel; and Dr. Mohan G. Hebsur for the NbYSi matrix material.

REFERENCES

1 HITEMP Review 1988. Advanced High Temperature Engine Materials Technology Program, NASA CP-10025, 1988.

2 HITEMP Review 1989. Advanced High Temperature Engine Materials Technology Program, NASA CP-10039, 1989.

3 HITEMP Review 1990. Advanced High Temperature Engine Materials Technology Program, NASA CP-10051, 1990.

4 HITEMP Review 1991. Advanced High Temperature Engine Materials Technology Program, NASA CP-10082, 1991.

5 McMaster, R. C., ed. *Nondestructive Testing Handbook. Vols. 1 and 2*. American Society for Nondestructive Testing, Columbus, OH, 1959.

6 Briggs, A. *Scanning Acoustic Microscopy*. Oxford Univ. Press, 1991.

7 Generazio, E. R., et al. "Nondestructive Evaluation of Fiber-to-Matrix Interface Shear Strength," to be published as NASA TM, 1995.

8 Welch, C. S., et al. "Material Property Measurements with Post-Processed Thermal Image Data," *Thermosense XII: An International Conference on Thermal Sensing and Imaging Diagnostic Applications*, SPIE, Vol. 1313, SPIE, Bellingham, WA, 1990, pp. 124–133.

9 Heath, D. M., et al. "Quantitative Thermal Diffusivity Measurements of Composites. Review of Progress in Quantitative Nondestructive Evaluation," Vol. 5B: *Proceedings of the Twelfth Annual Review*, D. O. Thompson and D. E. Chimenti, eds., Plenum Press, New York, 1986, pp. 1125–1132.

10 Hung, Y. Y. "Shearography: A Novel and Practical Approach for Nondestructive Testing," *J. Nondestr. Eval.*, Vol. 8, No. 1, 1989, pp. 55–67.

11 Briggs, A. *An Introduction to Scanning Acoustic Microscopy*. Oxford Univ. Press, 1985.

12 Cargill, G. S. "Ultrasonic Imaging in Scanning Electron Microscopy," *Nature*, Vol. 286, No. 5774, 1980, pp. 691–693.

13 Cantrell, J. H. and Quian, M. "Scanning Electron Acoustic Microscopy of SiC Particles in Metal Matrix Composites," *Mater. Sci. Eng.*, Vol. A122, 1989, pp. 47–52.

14 Martz, H. E., et al. "Nuclear-Spectroscopy-Based, First-Generation, Computerized Tomography Scanners," *IEEE Trans. Nucl. Sci.*, Vol. 38, No. 2, 1991, pp. 623–635.

15 SCS-6 SiC Fiber. Textron Specialty Materials, Lowell, MA.

16 Generazio, E. R. "Ultrasonic and Radiographic Evaluation of Advanced Aerospace Materials: Ceramic Composites," NASA TM-102540, 1990.

17 Generazio, E. R. "Theory and Experimental Technique for Nondestructive Evaluation of Ceramic Composites," NASA TM-102561, 1990.

18 Generazio, E. R. and Swickard, S. M. "Ultrasonic and Radiographic Evaluation of Advanced Aerospace Materials: Ceramic Composites," HITEMP Review 1990: Advanced High Temperature Engine Materials Technology Program, NASA CP-10051, 1990, pp. 65-1, 65-7.

19 Generazio, E. R., Roth, D. J., and Baaklini, G. Y. "Acoustic Imaging of Subtle Porosity Variations in Ceramics," *Mater. Eval.*, Vol. 46, No. 10, 1988, pp. 1338–1343.

20 Generazio, E. R., Roth, D. J., and Stang, D. B. "Ultrasonic Imaging of Porosity Variations Produced during Sintering," *J. Am. Ceram. Soc.*, Vol. 72, No. 7, 1989, pp. 1282–1285.

21 Generazio, E. R., Stang, D. B., and Roth, D. J. "Dynamic Porosity Variations in Ceramics," NASA TM-101340, 1988.

22 Generazio, E. R., Roth, D. J., and Baaklini, G. Y. "Imaging Subtle Microstructural Variations in Ceramics with Precision Ultrasonic Velocity and Attenuation Measurements," NASA TM-100129, 1987.

23 Biberman, L. M. *Perception of Displayed Information.* Ch. 7, Plenum Press, New York, 1973.

24 Bhatt, R. "Influence of Interfacial Shear Strength on the Mechanical Properties of SiC Fiber Reinforced Reaction-Bonded Silicon Nitride Matrix Composites," NASA TM-102462, 1990.

25 Bhatt, H., Donaldson, K. Y., and Hasselman, D. P. H. "Role of the Interfacial Barrier in the Effective Thermal Diffusivity/Conductivity of SiC-Fiber-Reinforced Reaction Bonded Silicon Nitride," *J. Am. Ceram. Soc.*, Vol. 73, No. 2, 1990, pp. 312–316.

26 Eldridge, J. I. and Honecy, F. S. "Characterization of Interfacial Failure in SiC Reinforced Si₃N₄ Matrix Composite Material by Both Fiber Push-Out Testing and Auger Electron Spectroscopy," *J. Vac. Sci. Technol.*, Vol. A8, No. 3, 1990, pp. 2101–2106.

27 Baaklini, G. Y. and Bhatt, R. T. "Preliminary Monitoring of Damage Accumulation in SiC/RBSN," HITEMP Review 1992: Advanced High Temperature Engine Materials Technology Program, NASA CP-10082, pp. 54-1–54-13.

PART 4

TECHNOLOGY BENEFITS ESTIMATION

CHAPTER 11

Technology Benefit Estimator (T/BEST) for Aerospace Propulsion Systems

EDWARD R. GENERAZIO[1] and CHRISTOS C. CHAMIS[2]

OVERVIEW

A technology benefit estimator (T/BEST) system has been developed to provide a formal method to assess advanced technologies and quantify the benefit contributions for prioritization. T/BEST may be used to provide guidelines to identify and prioritize high payoff research areas, help manage research and limited resources, show the link between advanced concepts and the bottom line, i.e., accrued benefit and value, and to credibly communicate the benefits of research. An open-ended, modular approach is used to allow for modification and addition of both key and advanced technology modules. T/BEST has a hierarchical framework that yields varying levels of benefit estimation accuracy that are dependent on the degree of input detail available. This hierarchical feature permits rapid estimation of technology benefits even when the technology is at the conceptual stage. T/BEST's software framework, status, novice-to-expert operation, interface architecture, analysis module addition, and key analysis modules are discussed. Representative examples of T/BEST benefit analyses are shown.

INTRODUCTION

Over the past several years, the costs and risks of introducing dramatically new technologies into aerospace propulsion systems have been

[1]NASA Lewis Research Center, Cleveland, OH, U.S.A.; current address NASA Langley Research Center, Hampton, VA, U.S.A.
[2]NASA Lewis Research Center, OH, U.S.A

perceived to be very high, while other benefits, e.g., noise, emissions, weight, and reliability, are generally unknown. This perception and insufficient knowledge of benefits of new technologies is an effective barrier for introducing new technologies into propulsion systems. A good example of perceived high-risk technology is composite materials. Here, the manufacturing costs, durability, and material properties are often unknown until detailed analyses are performed. That is, an engineer cannot simply turn to a handbook of standards on composites to obtain the required material data for designing a composite component. The engineer needs to perform a detailed multidisciplinary analysis, including effects of fiber orientation and external loads, etc., on the engineered component. Also, since the durability is not represented in a handbook, a component life analysis needs to be done in conjunction with the multidisciplinary analysis. Here, the internal makeup of the component, the component geometry, and the component's life can greatly affect the life cycle costs of the component. It is costs such as these that need to be quantitatively determined before new technologies are introduced into aeropropulsion systems. The over-arching goal of this work is to credibly communicate the benefits of research in new technologies. A formal method is needed to assess technologies and quantify and prioritize benefit contributions.

Such a method can be used for providing guidelines to identify and prioritize high payoff research areas, help manage research and limited resources for greatest impact, and to show the link between advanced concepts and the bottom line, i.e., accrued benefit and value. Credible determination of benefits can only be obtained from an optimization, with respect to industry-specific objective functions, of multidisciplinary analyses that include analyses such as mission, engine cycle, weight, life, emissions, noise, manufacturing, and cost. The following work describes the technology benefit estimator (T/BEST) analysis simulation system that is being developed for credibly communicating the benefits of introducing new technologies into aerospace propulsion systems. Typical benefits needed to make investment decisions are range, speed, thrust, capacity, city pairs reached, component life, noise, emissions, specific fuel consumption, component and engine weights, precertification test, engine cost, direct operating cost, life cycle cost, manufacturing cost, development cost, risk development time, and return on investment.

APPROACH

A hierarchical, modular, open-ended approach has been used to interface a wide variety of disciplines: mission analysis, thermodynamic engine cycle analysis, engine sizing or flowpath analysis, weights analysis, structural analysis, cost analysis, and noise and emissions analyses. The integration and control of these software analyses is the technology benefit estimator (T/BEST) system. The hierarchical arrangement permits credible estimation of benefits, even when there is a limited amount of data. Here, a package of statistical correlations is relied upon for projecting both the interpolated and extrapolated data that is needed for completing a benefits analysis. Increasing hierarchical detail yields increased accuracy of the benefits (Figure 11.1) determined. The hierarchical details span the range from speed, capacity, and range of the first estimation level to the constituent components of the subcomponent of the twelfth estimation level (Figure 11.2). The open-ended feature allows for system growth, while the modular aspect permits easy addition, updating, and replacement of analysis modules (Figure 11.3). The T/BEST system is designed to be operated by beginner, intermediate, and expert users. The beginner has control of basic parameters, e.g., mission, range, component materials, and airfoil geometry selection from the component's library, while the expert has full control of all input details of each analysis module. Using this approach, an expert in structural or fluid analysis can observe the effects of the details of their respective analysis on a mission or engine cycle, etc., without being an expert in mission or engine cycle analysis. The converse is also true, where an expert in mission or engine cycle analysis can see the effects of their respective details on the structural and fluid analysis (life and efficiency) without being an expert in structural or fluid analysis. The overall system level benefits are available to all user levels.

OVERVIEW OF T/BEST

Figure 11.3 indicates a representative set of discipline modules within the T/BEST system. The T/BEST executive, a UNIX shell, controls the operation of the entire system. All critical data, e.g., specific fuel consumption, number of stages, subcomponent weight, stress levels at critical areas of the components, etc., are passed through a neutral file so that other modules can easily assess and share this data. Copious data, such as airfoil profile geometry, nodal pressure, and temperature loads, are passed via pointers that point to the appropriate files. The flow of data in T/BEST is graphically shown in Figure 11.4, and a section of the neutral file is shown in Figure 11.5. The figure of merit or objective function varies (Figure 11.6) with the problem being addressed; however, a typical objective function is shown in Figure 11.7. Here, the changes in specific fuel consumption, weight, material, and maintenance costs are combined with coefficients and normalization factors (the coefficients and normalization factors are usually proprietary). The security of the T/BEST system is

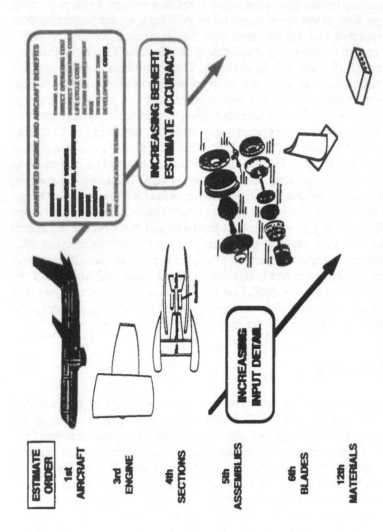

Figure 21.1 Hierarchical technology benefit estimator

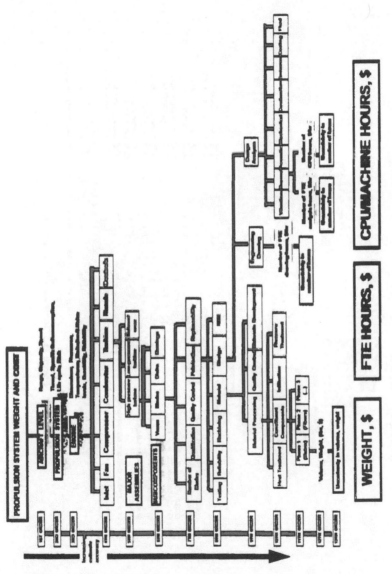

Figure 11.2 Technology Tree for aerospace propulsion system.

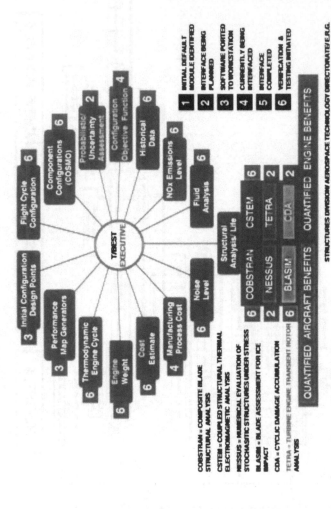

Figure 11.3 Technology benefit estimator.

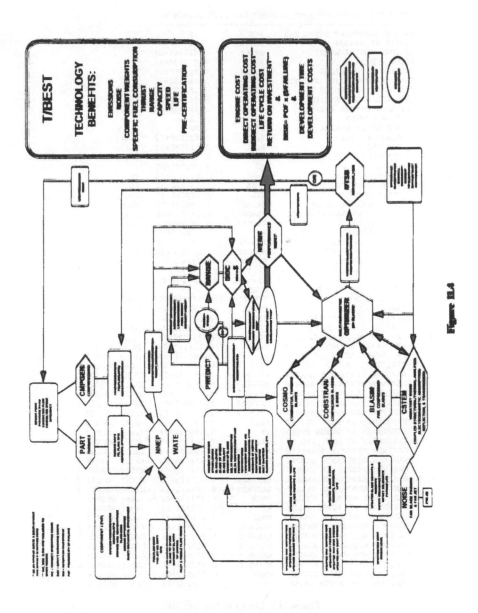

Figure II.4

211

```
*** TBEST EXECUTIVE SYSTEM - NEUTRAL FILE UPDATE ***
----------------------------------------------------
ENGINE COMPONENT TYPE: FAN    NCC        2
NUMBER OF STAGES              NSTAGE     2
MINIMUM CRUISE SPEED          RPMCR      0.61060E+04
ROTOR SPEED                   RPM        0.61060E+04
MAXIMUM ROTOR SPEED           RPMAX      0.61060E+04
BLADE TAPER RATIO (HUB/TIP)   TR         0.18000E+01
UPSTREAM HUB RADIUS           RIUP1      0.12000E+02   (in.)
DOWNSTREAM HUB RADIUS         RIDW1      0.24000E+02   (in.)
UPSTREAM SHROUD RADIUS        ROUP1      0.31000E+02   (in.)
  STAGE NUMBER                NS         1
  NUMBER OF BLADES            NB         33
  STAGE WEIGHT               NSTW        0.53000E+03   (lbs)
  HUB RADIUS                 RHBA        0.11770E+02   (in.)
  TIP RADIUS                 RTBA        0.30970E+02   (in.)
  ASPECT RATIO               AR          0.30000E+01
  MAXIMUM TEMPERATURE        TMAX        0.88400E+03   (R)
  BLADE ROOT ANGLE           THER        0.16433E+02   (deg.)
  STAGE LENGTH               STL         0.13900E+02   (in.)
  BLADE BROACH ANGLE         BRANG       0.00000E+00   (deg.)
  BLADE STAGGER ANGLE        STAGG       0.35000E+02   (deg.)
  1ST STATION CHORD LENGTH   CHORD(1)    0.82282E+01   (in.)
  STAGE PRESSURE RATIO       PR          0.20900E+01
  STAGE PRESSURE             STAGEP      0.19720E+04   (lb/ft^2)
  STAGE TEMPERATURE          STAGET      0.51900E+03   (R)
  STAGE MASS FLOW RATE       STAGEF      0.69390E+03   (lb/sec)
  BLADE MATERIAL             MATSLC      TITANIUM
  AIRFOIL DEFINITION         AIRCODE     NACA 64-206 FAN
  FULL BLADE DEFINITION      ABLDEF
  BLADE UNTWIST              UTWIST     -0.40361E+01   (deg.)
  BLADE UNCAMBER             UCAMB      -0.76460E+00   (deg.)
  MAXIMUM TIP EXTENSION      TIPX        0.33700E-01   (in.)
  MAX. IN PLANE Y-DISPL.     TIPY        0.11305E+01   (in.)
  MAX. IN PLANE Z-DISPL.     TIPZ        0.31400E-01   (in.)
  FREQUENCY AT MIN. CRUISE   WMC1        0.15116E+03   (cps) MODE 1
  FREQUENCY AT ROTOR SPEED   w1          0.15117E+03   (cps) MODE 1
  FREQUENCY AT MAX. SPEED    WRL1        0.15115E+03   (cps) MODE 1
  MAXIMUM RESONANCE MARGIN   MAXMR11     0.48536E+00   EXCIT. ORDER 1
  MAXIMUM RESONANCE MARGIN   MAXMR12     0.35732E+00   EXCIT. ORDER 2
  MAXIMUM RESONANCE MARGIN   MAXMR13     0.50488E+00   EXCIT. ORDER 3
  MAXIMUM RESONANCE MARGIN   MAXMR14     0.62865E+00   EXCIT. ORDER 4
  MAXIMUM RESONANCE MARGIN   MAXMR15     0.70293E+00   EXCIT. ORDER 5
  FREQUENCY AT MIN. CRUISE   WMC2        0.33270E+03   (cps) MODE 2
  FREQUENCY AT ROTOR SPEED   w2          0.33270E+03   (cps) MODE 2
  FREQUENCY AT MAX. SPEED    WRL2        0.33270E+03   (cps) MODE 2
  MAXIMUM RESONANCE MARGIN   MAXMR21     0.21709E+01   EXCIT. ORDER 1
  MAXIMUM RESONANCE MARGIN   MAXMR22     0.58547E+00   EXCIT. ORDER 2
  MAXIMUM RESONANCE MARGIN   MAXMR23     0.56982E-01   EXCIT. ORDER 3
  MAXIMUM RESONANCE MARGIN   MAXMR24     0.20726E+00   EXCIT. ORDER 4
  MAXIMUM RESONANCE MARGIN   MAXMR25     0.34581E+00   EXCIT. ORDER 5
  FREQUENCY AT MIN. CRUISE   WMC3        0.37537E+03   (cps) MODE 3
  FREQUENCY AT ROTOR SPEED   w3          0.37538E+03   (cps) MODE 3
  FREQUENCY AT MAX. SPEED    WRL3        0.37537E+03   (cps) MODE 3
  MAXIMUM RESONANCE MARGIN   MAXMR31     0.26885E+01   EXCIT. ORDER 1
  MAXIMUM RESONANCE MARGIN   MAXMR32     0.84427E+00   EXCIT. ORDER 2
  MAXIMUM RESONANCE MARGIN   MAXMR33     0.22981E+00   EXCIT. ORDER 3
  MAXIMUM RESONANCE MARGIN   MAXMR34     0.77866E-01   EXCIT. ORDER 4
  MAXIMUM RESONANCE MARGIN   MAXMR35     0.26229E+00   EXCIT. ORDER 5
  FREQUENCY AT MIN. CRUISE   WMC4        0.70356E+03   (cps) MODE 4
  FREQUENCY AT ROTOR SPEED   w4          0.70357E+03   (cps) MODE 4
```

Figure 11.5 Listing of "neutral.file."

212

T/BEST - MERIT

$$\text{MERIT}_I = \sum_J A_{IJ} * (X_{IJ}/N_{IJ})$$

A = PROPRIETARY COEFFICIENTS

N = NORMALIZATION FACTORS

X = DELTA T/BEST OUTPUTS (e.g., Δ WEIGHT, Δ COSTS, ETC.)

I = PERFORMANCE MERIT RELATIONSHIP NUMBER

J = NUMBER OF PARAMETERS IN RELATIONSHIP

$$\text{TOTAL MERIT} = \sum_I \text{MERIT}_I$$

Figure 11.6

ODJ = 0.43 * (% DELTA TSFC) + 0.40 * (DELTA WEIGHT/1000)
      + 0.21 * (DELTA COST/100,000) + 0.32 * (DELTA MC/10)

where DELTA TSFC = 6 * (1 - η) for rotor 6
                  = 60 * (1 - η) for fan
       η - Efficiency
       DELTA WEIGHT = No. of Blades * Blade Weight
       DELTA COST   = Material Cost
       DELTA MC   = Maintenance and Material Cost

Figure 11.7 Details of performance objective function (from T/STAEBL code).

213

assured, in that the entire system is transportable to a work station. Since the analysis codes are all in portable FORTRAN and UNIX shell, the transportability feature has come easily. It is pointed out here that software module developed using other languages, for example, object-oriented languages, can also be easily attached to T/BEST. After a complete simulation cycle or run, a wide range of data is available for display.

EXAMPLE

Figure 11.8 shows an initial engine configuration for a supersonic aircraft engine. During a simulation, T/BEST performs the thermodynamic engine cycle analysis, flight cycle optimization, engine sizing or flowpath determination and generates components and applies thermomechanical loads, structural analysis, weights analysis, fluid analysis, noise analysis, emission analysis, mean time between repair analysis, and direct operating cost analysis. Some typical outputs from T/BEST is shown in Figures 11.9–11.14. The blade weights, tip extensions, static blade root stress levels, uncamber and untwist, failure function, and rotor efficiencies are available stage-by-stage throughout the engine. Figures 11.9–11.14 show these results for the high-pressure compressor. The flight cycle mission is shown in Figure 11.15, and the direct operating cost is shown in Figure 11.16. The data in these figures and the constant design data in the next figure are all available from one simulation. Figure 11.17 indicates typical system-level benefits for constant design flow and constant design thrust designs where the combustor efficiency has been reduced by 10 percent. Parameter changes needed to perform evaluations take only a few seconds. Figure 11.18 indicates the increase of engine maintenance costs that occurs if the life of the combustor is modified. A complete simulation analysis that makes use of all of T/BEST's modules takes about twenty minutes. T/BEST may also be used for optimization. For example, optimization of the blade profile of the first fan stage, for minimized weight with tip thickness and resonance margin constraints, takes about forty-five minutes and eighty intracode cycles of the structural analysis code identified as BLA-SIM in Figure 11.4. Verification of the 120,000 lines of T/BEST software code is done as individual modules are attached and by independent evaluation by discipline experts.

CONCLUSIONS

A fast, credible, hierarchical, modular, open-ended technology benefits estimating system, T/BEST, has been developed and demonstrated.

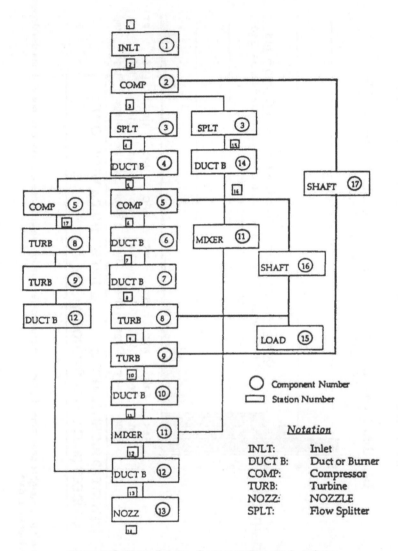

Figure 11.8 Block diagram of a supersonic research engine.

215

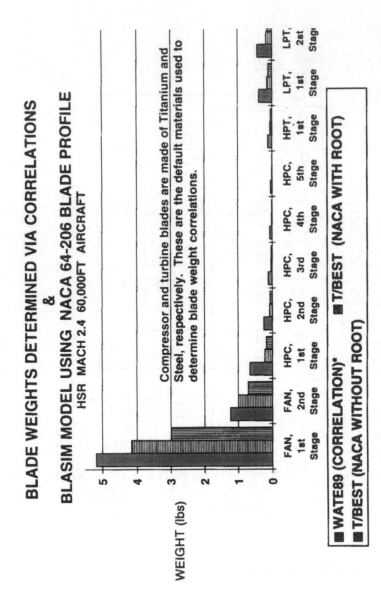

Figure 11.9 Blade weights comparison.

* The blade weights determined by the configuration evaluation weight code are based on correlations developed from historical data.

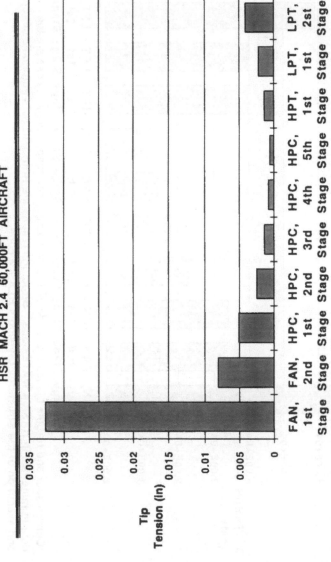

Figure 11.10 Blade structural response—tip extension.

*Compressor and turbine blades are made of Titanium and Steel, respectively. These are the default materials in NASA's WATE code. Structural results are from T/BEST's automatic set up and execution of the Structures Division's BLASIM code.

217

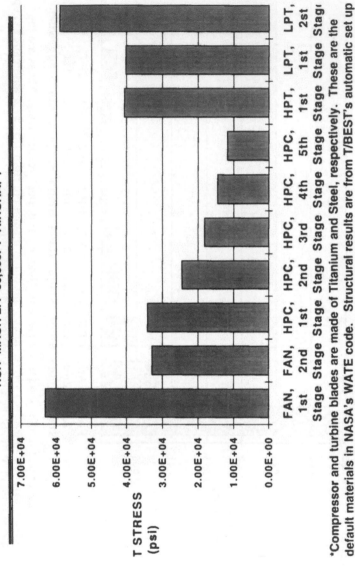

Figure 11.11 Blade structural response—root stress level.

*Compressor and turbine blades are made of Titanium and Steel, respectively. These are the default materials in NASA's WATE code. Structural results are from T/BEST's automatic set up and execution of the Structures Division's BLASIM code.

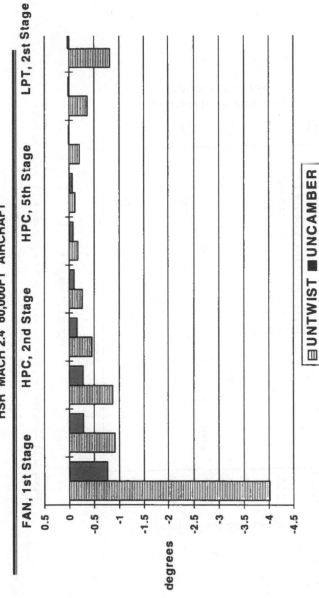

Figure 11.12 Blade structural response – UNTWIST and UNCAMBER.

*Compressor and turbine blades are made of Titanium and Steel, respectively. These are the default materials in NASA's WATE code. Structural results are from T/BEST's automatic set up and execution of the Structures Division's BLASIM code.

219

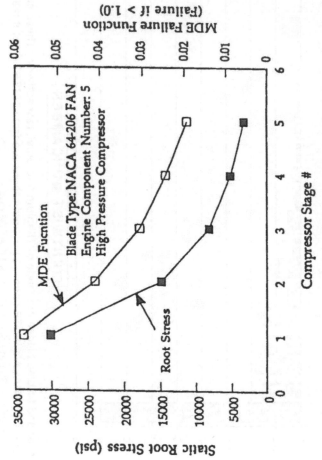

Figure 11.13 Blade structural response: static root stress level and MDE failure function.

220

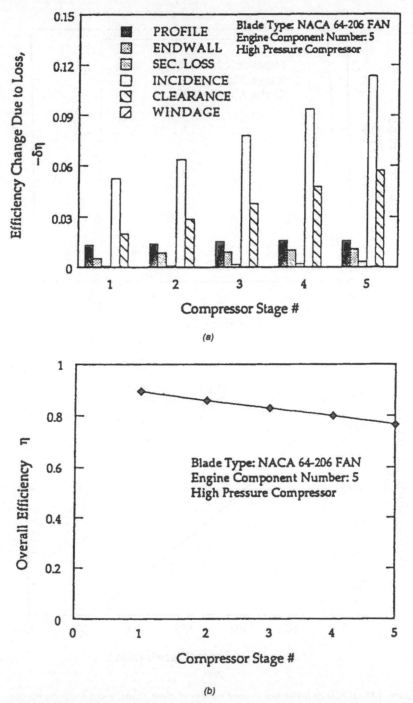

Figure 11.14 (a) Compressor efficiency changes due to loss; (b) high-pressure compressor stages – overall efficiency.

221

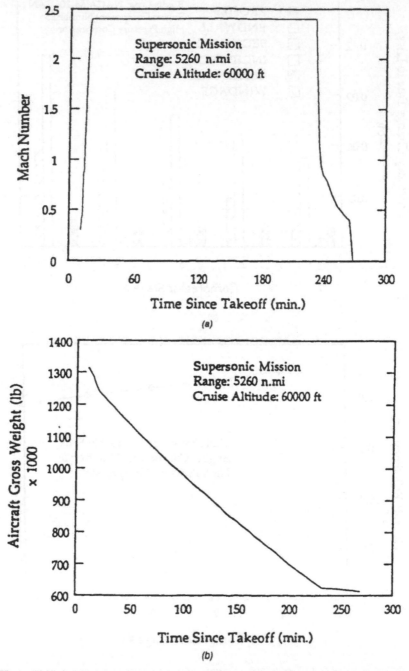

Figure 11.15 (a) Mission performance: mach number at climb, cruise, and descent; (b) mission performance: aircraft gross weight at climb, cruise, and descent.

222

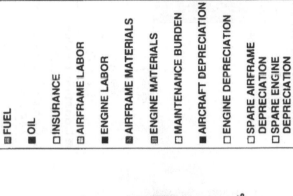

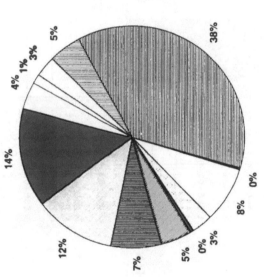

CAPTAIN/1ST OFFICER

FUEL

OIL

INSURANCE

AIRFRAME LABOR

ENGINE LABOR

AIRFRAME MATERIALS

ENGINE MATERIALS

MAINTENANCE BURDEN

AIRCRAFT DEPRECIATION

ENGINE DEPRECIATION

SPARE AIRFRAME DEPRECIATION

SPARE ENGINE DEPRECIATION

Figure 11.16 Turbine engine aircraft cost (%) mile @ design 60,000 ft.

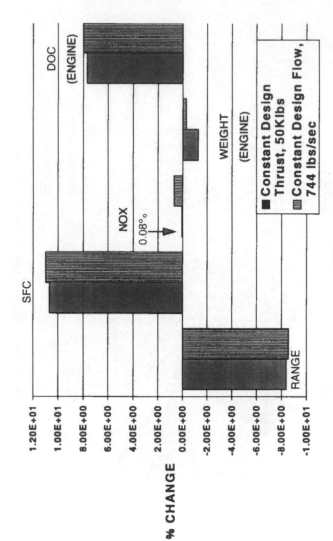

Figure 11.17 Combustor efficiency is 10% less than 0.99.

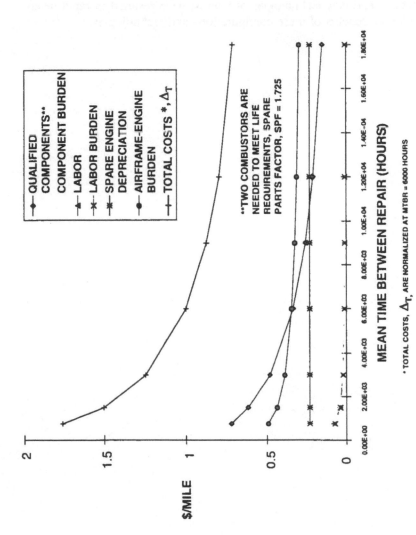

MEAN TIME BETWEEN REPAIR (HOURS)

*TOTAL COSTS, Δ_T, ARE NORMALIZED AT MTBR = 6000 HOURS

Figure 11.18 Engine maintenance costs.

225

T/BEST yields quantitative benefits of range, specific fuel consumption, emissions, noise, weights, thrust, capacity, speed, life, costs, precertification testing, etc. Initial applications of T/BEST to advanced engine configurations and new technologies of interest have resulted in rapid determination of benefits of these configurations and technologies.

T - #0528 - 101024 - C0 - 229/152/13 - PB - 9781566763349 - Gloss Lamination